Planning and Design of Engineering Systems

TITLES OF RELATED INTEREST

Analytical and computational methods in engineering rock mechanics
E. T. Brown (ed.)

Boundary element methods in elastodynamics
G. D. Manolis & D. E. Beskos

Boundary element methods in solid mechanics
S. L. Crouch & A. M. Starfield

Computers in construction planning and control
M. J. Jackson

A concise introduction to engineering economics
P. Cassimatis

Distribution-free tests
H. R. Neave & P. B. Worthington

The dynamics of rotor-bearing systems
M. Goodwin

Earth structures engineering
R. J. Mitchell

Engineering noise control
D. A. Bies & C. H. Hansen

Engineers in Britain
I. A. Glover & M. P. Kelly

The finite element method in thermomechanics
T.-R. Tsu

Hemispherical projection methods in rock mechanics
S. D. Priest

Hydraulics in civil engineering
A. Chadwick & J. Morfett

Intelligent planning
R. Wyatt

Marine geotechnics
H. Poulos

Numerical methods in engineering and science
G. deV. Davis

Plastic design
P. Zeman & H. M. Irvine

Rock mechanics for underground mining
B. H. G. Brady & E. T. Brown

Structural dynamics
H. M. Irvine

*Theory of vibration**
W. T. Thomson

* Originally published by, and available within North America from, Prentice-Hall.

Planning and Design of Engineering Systems

G. C. Dandy
R. F. Warner
Department of Civil Engineering, University of Adelaide

E & FN SPON
ALERE FLAMMAM
Taylor & Francis Group

First published by Unwin Hyman Ltd, 1989
Reprinted 2000
by E & F N Spon
11 New Fetter Lane, London, EC4P 4EE

E & F N Spon is an imprint of the Taylor & Francis Group

British Library Cataloguing in Publication Data

Dandy, G. C.
Planning and design of engineering systems.
1. Engineering. Design
I. Title II. Warner, R. F.
620′.00425
ISBN 0419 249400

Library of Congress Cataloguing-in-Publication Data

Dandy, G. C.
Planning and design of engineering systems.
Bibliography: p.
Includes index.
1. Engineering. 2. Engineering design.
I. Warner, R. F. (Robert F.) II. Title.
TA145.D235 1988 620 88–17353
ISBN 0 419 249400

Typeset in 10 on 11 point Times by Phoenix Photosetting, Chatham
Printed in the United Kingdom at the University Press, Cambridge

Preface

As engineering educators we have long felt the need to explain to first-year students the scope and nature of engineering, the type of work an engineer is called on to undertake, and, most importantly, the commonsense methodology which engineers adopt to solve real engineering problems. We believe our students deserve this explanation before they embark on the long (and sometimes tedious) analytic studies which form part of an engineering course.

Following a reorganization of the engineering course curriculum at the University of Adelaide, we were able to develop an introductory course specifically for our first-year civil engineering students. The subject, entitled Introduction to Engineering Planning and Design, has run successfully for seven years. Our lecture notes form part of the basis of this textbook.

We have found that our first-year introduction to engineering planning and design leads naturally on to a detailed two-semester undergraduate treatment of systems planning and design, including topics such as economic evaluation of engineering projects, social and environmental aspects of planning and design, and engineering decision-making. The present textbook thus contains material which can be covered in a two or three semester treatment of engineering planning and design.

The material in Chapters 1–4 provides the focal point for a one-semester introduction to engineering. In our first-semester subject we also include a brief introductory exposure to some of the concepts in Chapters 5, 6, and 9.

Obviously, an introduction to engineering can also lead into a study of structural design. Although we have not chosen that particular path, we see the introductory material in the first four chapters of this book as a convenient introduction to the structural strand in any civil engineering course. Indeed, we see significant advantages in using a first-semester introduction to engineering as an integrated starting point for the various synthetic and design-based strands in the engineering course.

As part of a first-semester introduction to engineering, we strongly recommend that students be actively involved in using and extending the information on methodology and creativity, rather than being exposed to it passively in the form of lecture material. In our teaching programme, students are required to come to grips with a realistic engineering problem during their first semester. Typically, they work in teams of five or six on the problem formulation and feasibility study phases of an engineering project and present a report at the end of a prescribed period of time. They receive limited help from a staff member who acts as adviser to the group.

Clearly, care has to be exercised in choosing project topics for students who have only a high-school knowledge of mathematics and science. On the other hand, one should not underestimate the creativity and ingenuity of young students whose creative insight has not been blinkered by years of studying closed-form solutions to analytic problems. Typical projects have dealt with topics such as the design of a movable cricket pitch, the design of a bushfire-resistant house, investigating the use of icebergs to supply water to Australia, the design of a nuclear fallout shelter, and the planning of flood protection works.

The second-semester subject is taught in the third year of our four-year degree course and it is based on Chapters 5–9 of the book. In Chapter 5 we cover some basic economic concepts and methods for economic evaluation of engineering projects. The social and environmental effects of engineering work are of increasing concern to professional engineers. These are considered in Chapter 6. Mathematical optimization techniques and their applications to engineering problems are dealt with in Chapters 7 and 8. Finally, Chapter 9 presents an introduction to engineering decision-making under risk and uncertainty.

Although the emphasis in this book tends to be on examples drawn from civil engineering, we believe the text is potentially of interest in other engineering fields.

It is not possible to mention here all of our colleagues, both in Adelaide and elsewhere, who have helped in the preparation of this book. However, we wish to thank Joerg Schneider, ETH Zürich, and George Sved and John Hatch of the University of Adelaide for their comments and criticisms of early drafts of various chapters.

While we are convinced of the need for a text which provides for students a useful and interesting introduction to engineering and to planning and design, we recognize that we have probably made various errors of omission and of fact, as well as simple mistakes. We would be grateful to receive comments, constructive criticism, and suggestions from users of this book.

G. C. DANDY and R. F. WARNER
Adelaide, April 1988

Contents

List of Tables

To Glenda and Sigrid

1 Nature and scope of engineering

In this introductory chapter we discuss the nature and scope of engineering work and the role of the engineer in the community. In broad terms, engineering is concerned with improving the quality of human life by maintaining and developing the complex physical infrastructure which is necessary for the functioning of modern society.

Engineers make extensive use of scientific and mathematical knowledge; however, their work is distinguished from the work of scientists and mathematicians by an emphasis on the practical use of knowledge to solve physical problems as economically and as efficiently as possible. Engineering work typically involves the planning of projects, the innovative design of engineering systems, practical problem-solving, and decision-making.

1.1 Society and its engineering infrastructure

Modern society functions within the framework of a physical infrastructure which is vast, complex, and pervasive. This infrastructure has been carefully engineered to supply water, power, transport, communications, and a myriad of other services, materials, and manufactured goods which are indispensable to modern everyday life.

At the most elementary level, the essential needs of any community are for shelter, food, water, energy, a healthy and safe environment, and protection from natural hazard and external attack. However, as society develops and becomes more complex, its infrastructure requirements also expand to include more sophisticated items such as transportation, telecommunications, a wide range of manufactured goods and exotic primary products, and the many technological adjuncts which are used in both work and leisure activities.

The engineering infrastructure has developed progressively over thousands of years in response to the expanding needs of society. However, society has itself evolved in the context of the available engineering infrastructure. Thus, the infrastructure is interwoven with the fabric of the society it supports. Indeed, the continuing history of science and technology is intimately associated with the history of civilization, and it is hardly meaningful to separate the one from the other. While engineering developments usually arise out of perceived community needs, they also have a feedback influence on society and can bring about profound changes in our way of life. Engineering-induced changes in society can be as significant and far-reaching as the more obvious ones brought about by political, social, and commercial developments.

For example, the transportation system in a city tends to grow in extent and capacity to meet the needs of the inhabitants. However, the efficiency of the transportation system in turn places limits on the physical size and the population of the city. Existing transportation technology is thus an important factor which determines the limits of city development. Furthermore, the efficiency and adequacy (or inadequacy) of the transport facilities profoundly affect the way of life of the city population.

In ancient city states such as Athens transport was severely limited, with the available technology based on the energy stored in human and animal muscles. This in turn limited the city to a size which allowed citizens to walk comfortably to all parts of the city and back in a day. The same limitation applied to imperial Rome and also to the medieval European cities.

It was not until the seventeenth and eighteenth centuries that improved transportation, based on fast horse-drawn vehicles, allowed cities such as London and Paris to develop in size and population. Only with the advent of steam trains and city and suburban railways in the nineteenth century did the large city with a commuting work force become feasible. In the twentieth century the internal combustion engine and the automobile have been responsible for the development of city conglomerates such as Los Angeles, with an extremely large population spread over an enormous area and with a life-style differing markedly from that in cities of traditional European size and form.

Engineering infrastructure in its modern format consists in part of large, interacting systems which are designed and operated to satisfy specific community needs. For example, the need for clean, fresh water in a city is satisfied by an extensive water supply system in which water is first collected in remote catchment areas and stored in dams, before it is transported by pipes and pumping stations to holding reservoirs close to the city. There it is filtered and treated before being reticulated through a complex network of pipes for industrial and domestic use in accordance with seasonal and daily demands.

Engineering systems which provide the so-called public services tend to be large, complex, and very expensive. A single system may encompass a number of different industries and government instrumentalities and provide work for many thousands of people. For example, the power consumed by an industrial city may well involve an industry in the search for new sources of fossil fuels, and another industry in the researching of improved production techniques and alternative sources of energy. An entire government department, with the assistance of various private companies, may be responsible for operating power stations to generate electrical energy, and for distributing that electrical energy. Other semi-government organizations are likely to be involved with environmental protection, resource management, and the maintenance of safety standards.

Of course, the engineering infrastructure does not consist solely of large-scale systems which provide public services. There is also a vast array of engineered installations of smaller scale which serve the needs of individuals, small groups of people, and organizations within the community. For example, television sets, apartment buildings, and chemical plants are engineered to satisfy the needs of individuals, families, and commercial enterprises, respectively.

The needs of individuals in society are catered for by the manufacture of large numbers of identical or nearly-identical products, such as kitchen utensils, refrigerators, radios, clothes, and automobiles. Although the size and cost of any one such consumer item may be relatively small, the total production is related to the size of the population and is often huge. Various engineering-based industries are thus involved in the operations of manufacturing and mass production.

1.2 Historical note

The terms **engineer** and **engineering** and their equivalents in other European languages appear to be of relatively recent origin, dating from the seventeenth century. Nevertheless, an engineered infrastructure has always been part of civilized society, and in this sense engineering is as old as civilization itself. It has thus been suggested that bridge engineering commenced when the first tree was felled across a brook too deep to ford, and that hydraulic engineering started with the deepening of a rivulet to improve the water supply for agricultural purposes (Watson 1890).

Engineering works of great complexity and large magnitude were undertaken in ancient Egypt and Mesopotamia. These included irrigation projects, road construction, and protective fortifications for cities. As early as the fourth millennium BC there was a government official in Egypt with the title Superintendent of Works, who was in charge of the construction and maintenance of public buildings and roads.

One of the earliest architect-engineers to be known by name was the Egyptian Imhotep, who is thought to have conceived and designed the first pyramids, including the stepped pyramid of Saqqara, near Memphis. Imhotep is also remembered for his medical expertise. After his death he was worshipped in Egypt as a deity.

The construction of the pyramids was one of the major engineering feats of the ancient world. It is not widely appreciated that the construction technique for the Egyptian pyramids involved much more than the simple placement of massive stone blocks on top of each other. A careful inward orientation of the surface elements was used to prevent stability failure and give the structure almost unlimited life. This has been responsible for the preservation of these massive structures over thousands of years.

The siege engine, a movable tower used to attack fortified walls, first made its appearance in Mesopotamia. It remained a feature of offensive warfare in the western world up to the appearance of the cannon. We shall see shortly that the term 'engineer' may well have been introduced in connection with a Roman version of the siege engine.

Great advances were made in pure science in the Greek city states in the first millennium BC. One of the first texts on mechanics, sometimes attributed to Aristotle but more often to Straton of Lampsakos, deals with static principles, levers, and gears, and stems from this period. A text on mechanics was also written by Archimedes. In the second and first centuries BC, mechanical devices which employed compressed air and water pressure were invented in Alexandria. These included water clocks and organs, a steam turbine, and compressed-air guns. The devices were little more than toys, which were used for entertainment. Unfortunately, the underlying concepts were not developed further at that time. Although great advances in science, mathematics, and architecture were made by the Greeks, engineering developments were not so apparent. This is sometimes attributed to an attitude held by many Greek philosophers, including Plato and even Archimedes, that knowledge should be acquired for its own sake and not for practical application.

Techniques of military and civilian engineering were taken over by the Romans, who developed and applied them to an unprecedented degree. The scale of achievement of the Roman engineers in the construction of roads, bridges, aqueducts, baths, and public buildings throughout the ancient world was in many cases not equalled again until the twentieth century. Their works remain most impressive even by today's standards of accomplishment. Remains of Roman civilian constructions and roads are still to be found in Europe and throughout the limits of the Empire.

Water was supplied to Rome through eleven aqueducts, the largest of which was over 100 km in length, with the water directed not only over the familiar multi-level arched aqueducts which traversed the plains, but also through extensive systems of tunnels. Water was reticulated in the city by means of pipes. It has been speculated that the use of lead waterpipes was responsible for extensive lead poisoning among the population of Rome before its fall in the fifth century AD.

Vast open-air auditoria, designed to hold tens of thousands of spectators (the Colosseum, for example, held more than 80 000 people), were supported on barrel vaults with extensive corridors, staircases, and rooms underneath. The construction in Verona is still in use for opera performances and concerts. Arches and domes were used to magnificent effect by the Roman architects and engineers; their work can be appreciated even today in buildings such as the Pantheon in Rome. Roman building construction was facilitated by the use of natural cements, which were obtained for example from the volcanic ash of Vesuvius.

Great emphasis was placed by the Romans on their roads, because these were the basic means of communication and allowed rapid deployment of troops throughout the territories. The roads were built at high initial cost to last indefinitely; some have lasted to the present day. The Roman philosophy of road design and construction is in sharp contrast with present-day concepts, whereby the design life of a pavement is often in the order of several years.

In the second century AD a Roman siege engine came into use which was known as an **ingenium**, i.e. an ingenious device. It was a form of battering ram used for attacking fortified walls. The operator of this machine, and of other military machines, came to be known as **ingeniator**. This is often considered to be the origin of the modern term **engineer**.

In the latter period of the Roman era, use was made of water power for milling and for sawing. These powered mechanical devices pointed the way to the future replacement of human and animal labour by other sources of energy, including water, wind, and steam.

Following the fall of Rome, the term 'engineer' came to be applied more and more to men who devised fortifications for cities. These were often the same people who developed methods of breaching the fortifications. Engineering textbooks written in Europe in the fifteenth and sixteenth centuries dealt extensively with the layout and construction of walled fortifications.

In this brief note it is not possible to do justice to the history of engineering development in Europe, let alone worldwide. Nevertheless, it must be emphasized that Europe was by no means the sole focus of developments in engineering and technology. Some of the devices and inventions already referred to had a parallel development in China and elsewhere. Chinese inventions such as the compass, the horse collar, and gunpowder had a profound effect when they were eventually introduced in Europe.

In Italy, at the end of the fifteenth century, Leonardo da Vinci produced plans and sketches for an incredible array of ingenious machines and devices for use in both military and civilian engineering. His proposals anticipated to a remarkable extent later developments in hydraulics, mechanics, mechanical and military engineering. Among his papers he left feasible plans for submarines, armoured tanks, multi-firing guns, automated looms, screw-cutting machines, and presses. Although some people were able to benefit from his work, the ideas which were produced by this genius did not become well known and unfortunately did not have an important influence on the development of engineering.

On the other hand, Galileo was for a time Professor of Physics and Military Engineering at the University of Padua. He published his 'Dialogues Concerning Two New Sciences', which dealt with the foundations of statics and dynamics. This might well be regarded as the beginning of engineering science.

The first formal school for military engineers was established in France in 1675. The Corps du Génie were trained by military engineers of the calibre of Vauban, whose star-shaped designs for city fortifications were well known throughout Europe. The school proved to be highly successful, and as a result a civil corps were formed in France in 1716, the Corps des Ingénieurs des Ponts et Chaussées. In 1747 the École des Ponts et Chaussées was established to train the Corps. In this school, which was at that time unique, the use of mathematics, geometry, mechanics (statics), and testing was regarded as the basis for engineering work of all kinds.

The French emphasis on the use of scientific principles was by no means universally followed. In the eighteenth century in England, engineering was largely an art which was learnt by spending an apprenticeship with a master. Among the successful, and indeed excellent, British engineers there tended to be a mistrust of mathematics, and theory was thought to be of little value to the practitioner. Thomas Telford, one of the supremely innovative and successful engineers of this time made distinguished contributions in the design of aqueducts, cast-iron bridges, harbours, canals, and earthworks, using, in his own words, 'no theory'.

In 1795, just after the French Revolution, Napoleon established the École Polytechnique in Paris. This set the pattern for the later establishment of polytechnics in Zürich, Delft, and many other European cities. The Massachusetts Institute of Technology, chartered in 1861, was the first technical institution of this type to be set up in North America and pioneered a number of engineering developments. In Scotland, and then in England, engineering appeared as a discipline in the traditional universities.

Modern engineering, with its emphasis on the use of scientific principles, developed gradually. Nevertheless, it is reasonable to identify its beginnings with the establishment of the École des Ponts et Chaussées in Paris. Certainly from this time there came from French academic engineers and scientists such as Navier, Cauchy, Coulomb, Poisson, Lamé, and Clapeyron a flood of significant developments in engineering science and new fields of application such as elasticity theory, thermodynamics, and materials science.

In France and also in Britain the term **civil engineering** came to be used in the eighteenth century to refer to non-military or civilian engineering. John Smeaton is thought to be the first person to refer to himself as a civil engineer. This first occurred in court proceedings when he was required to give expert evidence. In 1771, a group of engineers in England, including Smeaton, formed a society of engineers. In 1828 the society became the Institution of Civil Engineers. In its charter, engineering was defined as:

> . . . *the art of directing the great sources of power in nature for the use and convenience of man, as the means of production and of traffic in states, both for external and internal trade as applied in the construction of roads, bridges, aqueducts, canals, river navigation, and docks for international intercourse and exchange, and in the construction of ports, harbours, moles, breakwaters and light-houses, and in the art of navigation by artificial power for the purposes of commerce, and in the construction and application of machinery, and in the drainage of cities and towns.*

The American Society of Civil Engineers was instituted in 1852. Specialization within the ranks of the non-military engineers already began to occur in the early nineteenth century, with the development of steam power for factories and for locomotion. Indeed, the industrial developments of this period really got under way with the replacement of human and animal labour by other sources of energy, such as water and wind, and then steam, electricity, and solid and liquid fuels. A vast range of tools and machines were designed for use in the manufacture of heavy equipment for the railways and the textile industry.

In England, the Institution of Mechanical Engineers was formed in 1846 with Stephenson as its president. In the United States, the Society of Mechanical Engineers was formed in 1880. In the early part of the twentieth century the scope of mechanical engineering was extended enormously by the development of efficient and reliable internal combustion engines and their use in automobiles and aircraft.

The formation of the Society of Mechanical Engineers was actually preceded in the United States by that of the Institute of Mining and Metallurgical Engineers in 1871. The Institute of Electrical Engineers followed in 1884, and the Institute of Chemical Engineers in 1908.

Concepts of economic operation and cost control as part of the engineering process underwent significant development in the nineteenth century. Cost control and the efficient use of resources are always an important element in good engineering work; however, when new techniques are being developed and used for the first time, the prime objective is to complete the project successfully, irrespective of cost. The point has been made by Hicks (1977):

> *Until this time engineers were primarily battling the elements, and costs were assumed to be a necessary and relatively uncontrollable price for winning the battle against nature.*

Nowadays, cost control is of course an integral part of any modern engineering activity.

Although the telegraph was developed in the first half of the nineteenth century, the impetus for developments in electrical power engineering followed somewhat later from the inventions of Edison and Siemens. The first electric lighting system appeared in California in the 1880s, in conjunction with the commercial supply of electricity. The telephone also came into use at this time, following the work of Alexander Graham Bell. As the result of the pioneering work of Hertz, Marconi succeeded in the wireless transmission and reception of messages across the Atlantic. With the invention of the vacuum tube by de Forest in 1906, radio broadcasting was rapidly accomplished and by the 1920s had become a part of everyday life.

Following World War II, developments in all fields of engineering were rapid. An updated description of engineering was provided in 1958 in the United States by the Presidential Council on Science and Engineering:

> *Engineering is the profession in which a knowledge of mathematics and natural science gained by study, experience and practice, is applied with judgement to develop ways to utilize economically the materials and forces of nature for the benefit of mankind.*

Engineering developments have continued to occur at an ever-increasing pace in all fields of engineering, but particularly in electronics, communications, and information systems. All fields of engineering have been transformed in the past twenty years by the availability at low cost of powerful computing facilities.

However, there has been a developing concern in recent years with the over-use of resources and the need for careful management and conservation, rather than exploitation, of the environment. The battle has ceased to be *against* nature. Over-exploitation of water resources for irrigation has produced very serious salinity problems in many parts of the world; intensive use of automobilies for city transport has produced widespread problems of smog and toxicity; acid rain, caused by emissions from coal-burning power stations, has caused widespread damage to forests in Europe and North America.

It is noteworthy that over-use and depletion of natural resources and pollution are not uniquely modern phenomena. Deforestation in the middle ages caused vast and permanent changes to the landscape in parts of Europe. In England in the sixteenth century, Queen Elizabeth barred the felling of trees in certain regions where forests were being depleted to make charcoal. She also barred the use of soft coal in London because of the production of excessive amounts of smoke and soot.

Current problems of over-use of natural resources and stressing of the environment have focused the attention of engineers and of the community on the need for conservation and optimal long-term management of available resources. Ageing and deterioration of costly constructed facilities such as pipelines, sewers, city buildings, bridges, and highways have further emphasized the need for careful management and maintenance of the engineering infrastructure.

Current concepts of engineering have developed beyond the optimistic attitudes implied in the historical definition of the Institution of Civil Engineers quoted above. In the nineteenth century, nature was regarded by engineers and by society as an infinite source of power and natural resources which is available for exploitation and use. The need for economic use of resources was recognized in the 1958 definition. Today there is recognition of the fragility of nature and the environment.

More than ever before, engineering is concerned with community needs and with providing, maintaining, and improving the infrastructure of society. However, our perceptions of the needs of the community have changed: the emphasis has tended to shift away from the optimistic expansion and unlimited exploitation attempted in the nineteenth century towards maintenance and management of our assets and the preservation and husbandry of the earth's resources.

1.3 The traditional branches of engineering

At the beginning of the nineteenth century, there were only two traditional branches of engineering, namely **military** and non-military or **civilian** engineering. However, following rapid scientific developments and industrial expansion, non-military engineering developed enormously in scope, and specialization became necessary. By the end of the century five main branches of engineering had come to be recognized: **civil, mechanical, electrical, mining,** and **chemical** engineering. In the twentieth century, these have come to be regarded as the 'traditional' branches of engineering, as engineering work continues to expand at an unprecedented rate into new areas.

The scope of current engineering work is such that no one person can hope to master the detailed knowledge and technical activities of all the fields of modern engineering. Nevertheless, entry into any specific field usually takes place via tertiary study in one of the traditional branches. We shall consider briefly the scope of these traditional branches in turn, together with some of the newer fields which have derived from them.

Civil engineering

Civil engineering has traditionally been the branch of engineering responsible for large public works which provide facilities such as water supply, sewerage and drainage, public health, public buildings, and transportation. Public works engineering is involved with the design and construction of physical facilities and also with the efficient operation and maintenance of those facilities. Transportation engineering is thus concerned not only with the design and construction of roads, railways, canals, airports, and seaports but also with their effective use and operation to ensure that the movement of people, goods and materials is prompt and efficient. Civil engineering includes specialized fields such as water engineering, public health engineering, structural engineering, geotechnical engineering, construction engineering, and transportation engineering.

Water engineering, an important field of civil engineering, covers the study of rainfall and runoff and the hydrological cycle, and the dynamics of water moving in pipes and open channels, as well as the planning, design, and operation of water supply systems, irrigation systems, the design and construction of harbours, and the maintenance of coasts.

Public health engineering has traditionally been involved with water supply and the collection and treatment of sewage and industrial wastes, but the field has extended to deal with problems of pollution control, environmental protection, and resource management.

Structural engineering deals with the analysis and design of buildings, bridges, towers, dams, and the like. In the design of large buildings, the structural engineer usually works in conjunction with an architect. In some countries, the contributions of architect and engineer are combined and provided by an architectural engineer. This is the case in Japan, where, owing to severe earthquake problems, architectural design and structural design are integrated.

Geotechnical engineering is closely related to structural engineering, in that it deals with the behaviour of soil and rock, especially with regard to the adequacy of foundations used to support buildings and other structures. The geotechnical engineer is also concerned with the design and construction of earth-fill dams, pavements, cuttings, and tunnels.

Construction engineering is concerned with the organization of machinery, materials, and labour for the construction of buildings, highways and the like, and the execution of large earthworks.

Mechanical engineering

Mechanical engineering and civil engineering separated into distinct branches in the nineteenth century, when steam power came into wide use and industry became more and more mechanized. Mechanical engineers then began to concentrate on the provision of power for industry, the design of machinery, and the development of manufacturing processes. Following on from these origins, mechanical engineering became closely involved with the use of heat and energy, and thus with the theoretical study and the practical engineering applications of thermodynamics. Other important interests of the mechanical engineer developed in the areas of the dynamics of mechanisms and the properties and behaviour of solid and fluid materials.

Mechanical engineers deal with the design of engines, electric power generation turbines, machinery for manufacturing, and heavy equipment for the construction industry and for agriculture. They are responsible for the design and manufacture of household appliances and automobiles, as well as with providing air conditioning, refrigeration, and the control of indoor environments.

The organization and management of industrial and manufacturing processes is an important part of mechanical engineering which is covered by industrial engineers.

More recently, mechanical engineers have become involved with the automation of mechanical and manufacturing processes. With the availability of cheap, electronic, miniaturized control systems and their application to mechanical processes, there have been rapid developments in the new fields of robotics and mechanical–electronic systems (mechatronics).

Electrical engineering

Electrical energy came into widespread use towards the end of the nineteenth century with the introduction of electric light and electrical machinery. Electrical engineering was originally concerned with the problems of generating and distributing electrical energy in large quantities and devising means for converting this into other forms of energy. Power engineering has remained one of the main areas of interest of the electrical engineer. Electrical engineers, in conjunction with mechanical engineers, are responsible for providing and maintaining services such as lighting and electrical energy, and designing and manufacturing devices such as motors and generators.

However, with the development of low-energy devices such as the vacuum tube, and more recently the transistor and the integrated circuit, a vast new field of electronic engineering and microelectronics has developed. This field has expanded enormously since World War II, largely as the result of developments in the fields of computing and communications.

Communication engineering deals not only with telephone, radio, telex, facsimile, and television services but also with remote sensing, signal processing, fibre optics, and laser technology. Control systems involving both large computers and small microprocessors are daily finding new engineering applications in our modern society.

Mining engineering

The systematic mining and working of metals have been central activities in all civilizations since the Stone Age. In a real sense, mining engineering has played a key role in the development of both military and civilian engineering. More recently, mining and metallurgical engineering have provided an essential basis for advances in civil, mechanical, and electrical engineering.

Traditionally, the mining engineer is responsible for the exploration and location of mineral deposits and for the design, construction, and operation of mining facilities. The metallurgist deals with the processing of the mined ores to extract the required metals and other materials, and to process them.

The techniques of mining exploration have changed greatly in the past few decades with the introduction of sophisticated detection and sensing procedures such as those based on satellite technology and seismic equipment. The interests of the mining engineer have likewise broadened to include increasingly rare minerals and materials needed for the production of metals and ceramics for modern industry.

Petroleum engineering, closely allied with mining engineering, is involved in the search for fields of crude petroleum and natural gas, and with the extraction and transport of these materials. The refining and processing of petroleum and natural gas is undertaken by the chemical engineer, among other things.

Chemical engineering

Chemical engineering was originally concerned with the design, construction and operation of plants which bring about chemical reactions on a large (i.e. factory) scale. Beside the traditional fields of interest of chemical production in the sugar, paper, and petroleum industries, chemical engineers deal with energy production, the production of pharmaceuticals, the treatment of air and water and the general problems of pollution control.

When the chemical plant has been designed and constructed, the work of the chemical engineer is by no means over. It is also important to adapt and adjust the production process until an efficient, optimal operating procedure is achieved. Maintenance, repair, and upgrading of plant and processes also have to be carried out by the chemical engineer.

Although the production of semiconductors and silicon chips is usually regarded as the province of the electronics engineer, many of the production problems fall into the ambit of chemical engineering.

Recent developments in chemical engineering have derived from the application of biochemical principles, in particular the use of enzyme and microbial activity in applications as diverse as the processing of minerals and other raw materials and the production of hormones, antibiotics, and vitamins.

1.4 Modern developments

In the earlier part of the twentieth century, various fields of engineering specialization developed, such as aeronautical, automotive, agricultural, and industrial engineering. Specialization has continued, as the result of engineering applications being found for new fields of scientific knowledge. There has also been a general acceleration of pace in engineering developments since World War II in response to the needs of society, with numerous new fields of engineering coming into being each year. Fields such as biomedical, nuclear, aerospace, reservoir, systems, cryogenic, geomechanical, refrigeration, air conditioning, and computer hardware and software engineering are typical of postwar developments. One way to observe the vast scope of modern-day engineering is to scan in a technical library the scholarly journals which are published on a regular basis and devoted to specialist engineering fields.

Interdisciplinary engineering work

The above paragraphs tend to highlight increasing specialization and distinctions among both the traditional branches of engineering and the branches which have more recently developed. In practice, the differences will not always be clear cut. In fact, the various branches and fields of engineering overlap extensively and naturally. In large engineering projects, engineers trained in different branches work together in teams to solve engineering problems and to produce integrated designs and plans.

For example, pollution control is a field which requires contributions from civil, chemical, and mechanical engineers. The new, developing fields of robotics and mechatronics involve contributions from both mechanical and electronics engineers. Engineering projects which deal with large-scale community problems are almost always carried out by teams of engineers with representatives from the main branches of engineering as well as members of other professions such as economists and architects, so that proper use can be made of all relevant knowledge and expertise.

1.5 The nature and scope of engineering work

Although engineering work has tended to become progressively more specialized, even the most diverse fields of engineering have various common characteristics which can be used to identify the basic nature of engineering work. For example, a common characteristic of modern engineering work is the use of scientific knowledge and mathematics wherever possible. Also, there are various activities which are common to the different fields and branches of engineering. These include planning and design, analysis, implementation, research and development, project management, and sales and marketing of engineering products and services. Taken together, these activities provide a good indication of the nature and scope of engineering work. We therefore consider them briefly in relation to the conduct of an engineering project. It is useful to begin with a discussion of engineering projects.

Engineering projects

In broad terms, engineers develop, maintain, and improve the physical infrastructure, and thereby improve the conditions of life in the community. Because of the unlimited needs and demands of the community on the one hand, and the limited availability of resources such as manpower, energy, materials, and money, on the other, it is essential that engineering work be carried out efficiently, with minimum use of scarce resources, and in limited time. For this reason, modern engineering work is usually undertaken in the form of goal-directed projects.

Typical examples of engineering projects include the design, construction, and operation of a chemical plant to produce fertilizer, the design and manufacture of a clutch system for a new model of automobile, the adaptation and programming of a microprocessor for the operation of a programmed washing machine, and the widening and strengthening of a bridge in order to allow the widening of a highway. A large part of modern engineering work is also concerned with the maintenance and upkeep of existing facilities. Although minor maintenance work may be undertaken by non-technical staff on a routine basis, any substantial maintenance and repair work needs to be executed as a carefully planned engineering project.

A large-scale engineering project requires the services of many engineers, usually drawn from a number of separate organizations and with a variety of different areas of expertise, possibly working with large numbers of technicians, tradesmen, and manual workers. On the other hand, a project with a modest goal, such as the design and construction of a small factory building, may be adequately handled by one or several engineers working with a small number of technical assistants and a small construction team.

Irrespective of the scale of an engineering project, the level of complexity is such that it has to be broken down into a number of tasks and sub-tasks which can be handled separately. These usually group themselves naturally into two main phases, the first being the **planning and design phase** and the second the **implementation phase**. Careful attention must be paid to the overall management of the project, so that the tasks and sub-tasks integrate properly and the work can be carried out efficiently and in a cost-effective manner.

Planning and design

Planning and design are activities in which the details of an engineering system or a process are determined to the extent necessary to allow implementation to be undertaken. For example, before work can start on the manufacture of a device or the construction of a building, the design of the device or building has to be carried out, and then the processes of manufacture or building have to be carefully planned. The initial planning and design phase in an engineering project is crucially important, and usually lengthy.

The terms **planning** and **design** are not mutually exclusive and there is no consistency in the technical usage of these words. Nevertheless, the tendency in this book will be to refer to the design of physical systems and components, and to the planning of operations and processes, i.e. to the **design of hardware** and the **planning of software**. According to this usage, one refers to the design of a bridge, and to the planning of the process of bridge construction. Design work is thus undertaken with the purpose of determining precise details of any devices, machines, or physical systems which are to be constructed or manufactured as part of the project, while the planning work is undertaken to determine the details of the procedures and processes which form part of the project.

For example, in the case of a project which has as its goal the provision of a new dam to extend the capacity of an existing water supply system, initial planning work is required to determine how the new dam will fit into the existing system and to determine its required capacity and possible locations. Detailed design work is carried out to establish the size and shape of the dam wall, and how it is to be keyed into the sides and floor of the valley. Design work is also carried out for the pipes, pumping station, filtration plant, and other items required to transfer water from the dam into the water supply system. Extensive planning is needed to decide on the construction sequence, on the specific steps involved, and how and when these steps are to be carried out. The construction steps may include provision of access roads, bringing in of heavy equipment to the site, excavation of soil and weathered rock, shaping and preparation of the rock foundations, grouting of unsound parts of the foundations, construction of main wall in segments, and so on.

For the mass production of a device such as an air conditioning unit, both planning and design are required. The physical device must be designed. Initially a good deal of planning activity is required to produce a design specification and design criteria to ensure that the device will function in the desired manner in conjunction with other engineering components. The specification should also take account of the perceived need for the product, so that there will be a good chance of commercial success. The physical details of the device are then determined in a design phase, while details of the manufacturing and assembling processes, including supply of parts, sequencing of factory procedures, and supply of items to end users, including sales and marketing as necessary, all have to be carefully planned.

Planning and design are crucial activities in the conduct of any engineering project. The quality of this work inevitably has a decisive effect on the success of the project. Planning and design are required in almost every type of engineering project, irrespective of the branches which may be involved. A large part of this book is devoted to the question of how the basic engineering activities of planning and design are undertaken.

Planning and design are also complex activities; so complex, that they can only be carried out in an iterative manner, with much trial and error.

Implementation

The major costs of an engineering project are usually incurred in the implementation phase. Implementation may involve extensive construction operations in the case of a dam or new chemical plant or factory. Manufacturing, with the prior installation of equipment and plant, may form the main part of the implementation phase of some mechanical and electrical engineering projects. In yet other projects, the development of efficient plant operation procedures may be of prime importance. Various specialized branches of engineering have evolved which deal specifically with the project implementation phase. These include construction management, industrial engineering, manufacturing engineering, methods engineering, and so on. Much detailed engineering knowledge and many specialized techniques are incorporated in these branches of engineering, and are dealt with in appropriate undergraduate courses and textbooks.

Engineering management

Engineering management is concerned with the planning and organization of an engineering project in order to ensure that it can be undertaken efficiently and carried successfully through all stages to completion. As the scale and complexity of a project increase, so too does the need for careful and effective planning and management. The objective of project management is not only to carry the project through to completion, but to do this efficiently in reasonable time and with minimum cost. The more complex the project, the more people and firms and groups will be involved, and hence the greater is the need for good management to ensure that the various groups work together harmoniously. Some of the aspects of an engineering project which require careful management include finances, material resources, equipment, human resources, time, and, not least, information.

It is normal practice for a project manager to be appointed to each engineering project. Depending on the scale of the work, use may be made of a project management team, with management expertise in a range of relevant fields. For example, in a multistorey building project the team would probably include structural, mechanical, electrical, and civil engineers, as well as architects, quantity surveyors, and accountants. The project manager and the management team are responsible for the smooth and successful progress of the work. The importance of engineering management in the implementation phase has already been emphasized; however, as the complexity and scale of the project increases, so too does the need for careful management of the planning and design phase.

Engineering analysis and modelling

Engineering analysis is concerned with the investigation, usually by means of theoretical calculation but possibly also by experimentation, of an existing or a proposed engineering system or component, in order to determine the way it will perform under a specific set of operating conditions. Full details of the system are known or assumed by the analyst.

Engineering analysis is undertaken as part of the design process in order to determine how a possible system design might function or otherwise perform in practice. The structural engineer, for example, analyses a tall building to determine the internal forces and stresses which would be produced by certain extreme wind or earthquake loads and gravity forces acting on the building and its contents. Likewise, the electrical engineer may analyse a circuit to determine currents and voltages in various components under typical conditions of use.

In some engineering fields, such as structural engineering and electrical engineering, there are well-formulated theories, based on physical principles, which allow engineering analysis to be undertaken on a routine basis, probably with the aid of existing computer software packages. In other situations, especially in some of the newer fields of engineering, the analytic procedures are not well developed, and the engineering analyst has to develop simplified mathematical models to suit each situation. The process of constructing an abstract, mathematical representation of a real world system is often called **modelling**.

Analysis, or mathematical modelling, is different in nature to the synthetic activies of planning and design, which are concerned with choosing appropriate details for a proposed new system or component or process. As already suggested, analysis may play an important part in the process of engineering planning and design. When, for example, the designer has tentatively chosen the details for a proposed device, the next step is to analyse the performance of this device under a wide range of operating conditions and thereby check that it will in fact perform in an acceptable manner.

It is not always possible to develop appropriate mathematical models or theoretical analyses in the limited time-frame in which design engineers have to work. Sometimes the physical conditions are far too complex for mathematical analysis, even with the assistance of the best computing facilities available. For example, in the study of the protection of a coastal region against erosion, the complex, interrelated three-dimensional movements of wind, water, and sand may best be handled by constructing a small-scale laboratory model.

Research and development

Basic research is often thought to be a typically scientific activity in which new knowledge is sought for its own sake. However, research is also an important engineering activity. Engineering research covers a spectrum ranging from basic research, with minimal concern for applications, through to the goal-directed investigation of a design engineer who must rapidly obtain new information, not currently available, in order to complete a design. Although engineering research extends across the full spectrum, it is more usually concerned with solving practical problems and has fairly specific goals.

Research and development work is often thought to be a one-way sequence of activities in which the results of pure research provide the potential basis for future engineering applications and development work. This is not always the case. Indeed, many major developments in both science and mathematics have originated in problem-oriented engineering investigations.

The value of applied research is recognized in all branches of engineering. In most engineering industries, a considerable effort goes into research and development work (R & D). The newest and most rapidly developing technologies also tend to be those where most money is invested in R & D. For example, in the microelectronics industry in the United States, approximately 7–10% of the annual budget for the industry is invested in research. Such investment pays for itself in profits from new products, and indeed is necessary for corporate survival in a rapidly changing industry.

Engineering R & D is carried out in universities, in the research laboratories of government and semi-government organizations, and in the laboratories of private companies. In some instances, notably the construction industry, the majority of R & D is undertaken with public funding in universities and government departments. On the other hand some industries, such as those involved in manufacturing, undertake a great deal of engineering research privately. If the results of such research are of great commercial value, they are likely to be kept secret for a period of time.

R & D includes testing of new materials and new products, the development of new production and manufacturing processes, and the development of new methods of engineering analysis. In brief, engineering research is concerned with the progressive development of new solutions to engineering problems.

Sales and marketing

While planning, design, and R & D are all clearly in the mainstream of engineering work, there are many additional related tasks which have to be carefully and thoroughly undertaken in order to ensure the success of an engineering project. In developed countries, sales and marketing are extremely important to the successful introduction of a new and innovative engineering product onto the market. The best engineering product will fail, irrespective of the quality of its design and manufacture, if it is not successfully brought into use.

It is not always appreciated that sales and marketing often require considerable engineering expertise. The more specialized and sophisticated a product is, from the engineering point of view, the more necessary it is that the sales and marketing personnel have a proper engineering background to appreciate the advantages of the product. An example is in the sale of large mainframe computers. A competent computer sales person needs a broad knowledge of the proposed fields of use for the computer, but also a deep knowledge of computer engineering and of the capabilities of the computer being sold, in relation to competing computers on the market. In the case of a complex engineering product, it is necessary for properly trained engineers to enter the fields of sales and marketing to ensure the success of that product.

1.6 Engineering, science, and mathematics

It has already been pointed out that, because of the unlimited demands which society makes on its limited resources of manpower, finances, materials, and energy, engineering projects have to be carried out efficiently at minimum cost, and in limited time. To achieve maximum efficiency, use is made of all available relevant knowledge. In particular, engineers draw wherever possible on scientific principles and mathematics. However, such knowledge is in itself insufficient, and usually has to be supplemented, for example by engineering R & D work, by experiment, and by knowledge gained from past experience.

Although engineers are users of scientific and mathematical knowledge, there are important features which distinguish the work of the engineer from that of the scientist and mathematician. The engineer has an active role to play in using resources to satisfy community needs. In contrast, the scientist tends to be more concerned with understanding and modelling the physical world, and is thus engaged in acquiring knowledge. The active work of the engineer is also distinct from that of the mathematician, who is concerned with logical structures which may or may not be observable in the physical world. While the prime engineering concern is with achievable community goals, and not with the discovery of new scientific or mathematical knowledge as such, good engineering work often leads to new scientific, mathematical, and engineering knowledge.

1.7 Engineering as creative problem-solving

Engineering is concerned with the solution of the problems which arise from community needs. These needs, and their related problems, are initially ill defined. Indeed, one of the first steps in engineering work is to identify the real problems which need to be solved. Even when the specific needs and problems have been clearly delineated, it is usually found that a number of alternative engineering solutions can be developed. Engineering problems rarely have unique solutions and, in this sense, are often described as being **open-ended**. Engineering work is often therefore described as **creative problem-solving**. Creativity and innovation are extremely important prerequisites for the successful handling of engineering problems, and it follows that the engineer needs to be creative and innovative in the search for effective solutions.

Much of the planning and design work which is undertaken in the initial phase of an engineering project falls into the category of creative problem-solving. However, creative problem-solving also makes up a large proportion of the work of implementation and management. The need for creative problem-solving in all stages of an engineering project cannot be over-emphasized. When an engineering project has been carefully planned, well conceived, and successfully implemented it can give the impression that the work was clear cut and inevitable. Such an impression can be very misleading: innovative solutions to open-ended problems appear to be inevitable only in retrospect.

One of the most difficult aspects of creative engineering work for students who have been trained in the class room to solve the typical, closed-form analytic problems of mathematics and physics, is to come to terms with real-world problems which do not have single 'correct' answers. In the next chapter of this book a methodology is described for planning and design which can also be used generally to undertake engineering problem-solving. One of the most important steps, after trying to clarify what the real problem is, involves the creation of a wide range of feasible alternatives. It is in this stage of the work that creativity and innovation play a crucial role. By evaluating and comparing the alternatives, the best solution can be identified and then implemented.

In summary, we may say that engineering work involves the innovative, creative, and efficient use of scarce resources to produce, maintain and operate the engineering systems which support modern society. The engineer makes use of available knowledge and experience in the search for creative and innovative solutions to the technical problems which arise out of community needs.

References

Asimow, M. 1962. *Introduction to design*. Englewood Cliffs, NJ: Prentice-Hall.

Armytage, W. H. G. 1976. *A social history of engineering*, 4th edn. London: Faber & Faber.

Beakley, G. C. & H. W. Leach 1972. *Engineering, an introduction to a creative profession*, 2nd edn. New York: Macmillan.

Cowan, H. J. 1971. *The master builders*. New York: Wiley.

Cowan, H. J. 1978. *Science and building*. New York: Wiley.

Hicks, P. E. 1977. *Introduction to industrial engineering and management science*. Tokyo: McGraw-Hill Kogakusha, International Student Edition.

Hollister, S. C. 1966. *Engineer*. New York: Macmillan.

Leavesley, J. H. 1987. *Mystique, magic and medicine*. Sydney: Australian Broadcasting Corporation.

Meredith, D. D., K. W. Wong, R. W. Woodhead & R. H. Wortman 1985. *Design and planning of engineering systems*, 2nd edn. Englewood Cliffs, NJ: Prentice-Hall.

Svensson, N. L. 1974. *An introduction to engineering design*. Randwick: New South Wales University Press.

Watson, J. E. 1890. The beginnings of engineering. *Transactions of the American Society of Civil Engineers*, **XXIV**. Reprinted in: *The civil engineer: his origins*. ASCE Historical Publication No 1 (1970).

Problems

1.1 Consider your activities throughout the course of a typical day, from the time you wake up in the morning until you go to sleep at night. Identify the parts of the engineering infrastructure that you rely on to undertake these activities.

1.2 Identify the various components of the engineering infrastructure which supports the city or region in which you live. Estimate the capital cost of replacement of this infrastructure. Estimate also the yearly running costs.

1.3 Consider the problem of establishing a small scientific base in Antarctica, to be permanently manned by 20 people. The base is to have shipping access, and will receive supplies once a year in December. Make a complete list of the components of the engineering infrastructure which will be needed to run the base successfully.

2 Engineering methodology

In this chapter a methodology, or working procedure, is described for undertaking the basic engineering activities of planning and design. The procedure relies on a commonsense, systematic approach, and is applicable to creative problem-solving generally. Creativity and innovation are important ingredients in the method.

2.1 The need for an engineering methodology

Engineering is largely concerned with solving complex and ill-defined community problems which by their nature do not have unique, 'correct' solutions. The work of the engineer has thus been described in the previous chapter as creative problem-solving. This is an apt description not only of the overall work of planning and design, but also of the detailed work which must be carried out in each phase of an engineering project.

Given the open-ended nature of engineering work, various questions arise regarding the way such work can best be undertaken:

- how should we go about solving an engineering problem, if there is not going to be a 'correct' solution?
- how can the planning and design work for an engineering project be best carried out?
- how can we decide whether a concept which has been developed in the planning and design phase of a project is the best available, or for that matter whether it is even acceptable?

An engineering methodology, or working procedure, is clearly needed. In the following section a commonsense approach for solving open-ended problems is described. The various steps in the approach are listed and will be discussed in some detail, specifically in relation to the activities of planning and design.

2.2 Problem-solving strategy

Engineering problems are inherently open-ended. They are usually also poorly formulated and ill defined, at least initially. There are good reasons for this. For example, the need for an engineering project may gradually come to be recognized in response to vaguely perceived community needs. Alternatively a project may develop through individuals or groups within the community who are likely to express their needs in non-technical and imprecise terms.

If an engineering problem is ill defined or poorly formulated, then common sense suggests that it should be clarified at the start in order to obtain a clear and unambiguous statement of the real problem.

Given that there is not likely to be a single 'correct' solution, even if the problem has been redefined in terms which are as precise as possible, it seems advisable *not* to search for just one solution, but rather to identify various alternative approaches, and indeed to make the range of possible solutions as broad and diverse as possible.

In order to choose from among such a range of alternatives, criteria have to be devised which in some way allow the alternatives to be compared and ranked. If this can be done then the preferred approach can be chosen and studied in greater detail and then implemented.

These brief comments contain the basis for a problem-solving strategy which may be summarized as follows:

- define the problem clearly
- generate a range of feasible approaches
- identify the best of these approaches
- work out the details for a solution, based on this approach
- implement this detailed solution

In the initial step of problem definition, or problem formulation, the aim is to clarify and if possible quantify the technical problem. Engineering problems involve many constraints which relate to costs, time, manpower, and materials. Proper account must be taken of all such constraints in formulating the problem. Criteria for evaluating, comparing, and ranking alternative approaches are often implied when a good problem statement has been found. In any case, the evaluation criteria need to be decided on in the initial problem formulation step. In most engineering projects, minimum financial cost is one of the most important of the criteria for evaluating alternatives. However, there are usually others.

In the second step, the task is to identify a wide range of alternative approaches, without overlooking any unusual, but potentially very effective, approaches. It is therefore advisable in the early stages not to concentrate on the first (usually the most obvious) approach which comes to mind, but rather to try to generate as wide a range of different and contrasting alternatives as possible. This increases the likelihood that an approach will be adopted which is, if not the best possible, at least effective and efficient, and better than most of the other possibilities. Some initial checking of the alternatives is also needed to ensure that they are feasible, i.e. would in fact solve the problem without violating the various constraints. Some culling of alternatives is thus undertaken, so that, at the end of the second step, a short list of good, feasible approaches has been identified. If no feasible approaches can be found, then the work on the problem has to be aborted or postponed. Alternatively, a redefinition of the problem may be attempted.

In the third step, a more detailed comparison is made of the alternative approaches remaining on the short list, with the purpose of identifying the 'best' one. The basis for the comparisons is the set of criteria developed in the initial problem formulation step. When the best alternative has been chosen, full details have to be worked out so that it can be implemented. This is the purpose of the fourth step. The final step is the implementation.

This procedure is a general one which can be applied to most open-ended problems. However, rather than continue the discussion in general terms, we shall consider this problem-solving strategy specifically in relation to engineering planning and design.

2.3 Methodology for planning and design

Figure 2.1 shows the two main phases of an engineering project, i.e. the initial phase of planning and design, and the final implementation phase. In Figure 2.2, the planning and design phase has been subdivided into four steps which correspond to the steps of the problem-solving procedure introduced in the previous section.

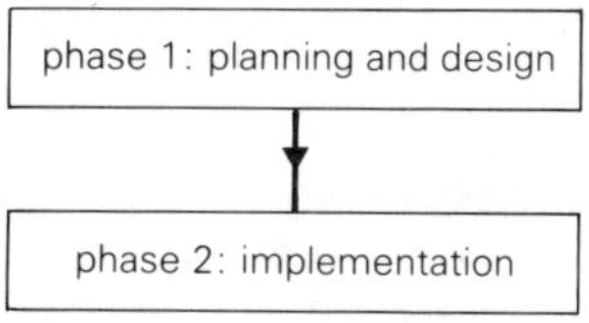

Figure 2.1 Two main phases of an engineering project.

In dealing specifically with planning and design, it is convenient to introduce a slight change in terminology, as indicated in Figure 2.2. In particular, the term **feasibility study** is used to refer to the step in which a wide range of workable (i.e. feasible) approaches are generated and evaluated. Some elementary analysis and planning and design work may need to be carried out for each alternative approach, so that a rough costing and evaluation can be made, as well as a check to ensure that the constraints of the problem are not violated. This is the stage at which 'back-of-the-envelope' calculations are used to investigate new concepts. Simple order-of-magnitude calculations and cost estimates are often adequate to identify new, promising ideas or to eliminate ineffective approaches. The non-feasible and non-competitive approaches are eliminated in the feasibility study.

In Figure 2.2, the third step in the procedure is referred to as **preliminary planning and design**. The purpose of this step is, as already stated, to identify the best of the feasible alternatives generated in the feasibility study. However, before such a choice can be made, additional information will usually be required on each candidate approach, so that more accurate costing and evaluation can be carried out. The initial design (or planning) work which was commenced in the feasibility study is thus developed and refined for each of the remaining approaches to allow the final choice of the best approach to be made.

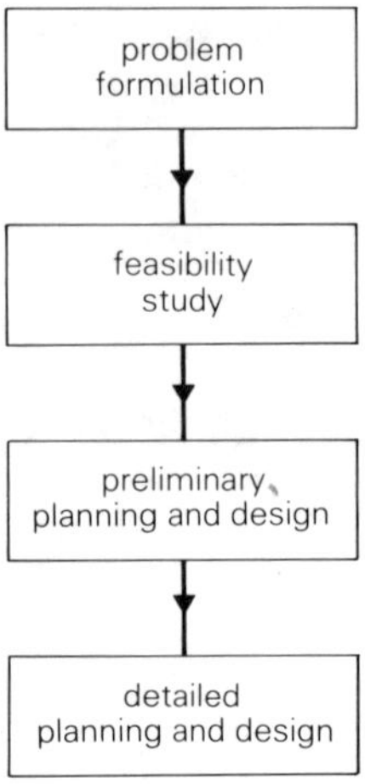

Figure 2.2 Steps in the planning and design phase.

In the fourth step, **detailed planning and design**, the best approach, already identified in the previous step, is studied carefully, in order to determine full details for implementation. Most of the complex calculations for engineering analysis are undertaken here.

It will be appreciated that the initial work of planning and design in the feasibility study is carried out with very simple and very approximate calculations for a wide range of alternatives. However, the calculations, cost estimates, and evaluations become progressively more complex and more accurate in the preliminary design and planning step, as the number of alternatives is reduced. This approach, of progressively increasing the accuracy and detail of the planning and design work as the number of remaining approaches decreases, leads to a reasonably effective use of engineering effort.

Although it is obviously advantageous to eliminate the less competitive approaches as early as possible in the process, it is sometimes found that two or more solutions are very nearly equal in cost, and are otherwise comparable. Very detailed calculations may then be required to choose the best approach. In such situations, full detailed designs may be completed for the alternatives, in order to allow the final decision to be made. This sometimes occurs in the design of large highway bridges where several structural concepts may be carried through the detailed stages of design before a final choice is made.

To carry out engineering planning and design work effectively it is advantageous to have available very simple methods of analysis for use in the feasibility study, when quick but inexpensive and approximate results suffice. More complex, more time-consuming, and more costly methods are used in the later detailed work, when greater accuracy is needed. There is thus a need for a range of analytic methods for use in design and planning, with varying levels of accuracy and corresponding variations in levels of complexity.

In Figure 2.3 the second main phase of an engineering project, i.e. implementation, is added to the steps in the planning and design phase to produce the overall view of the engineering methodology.

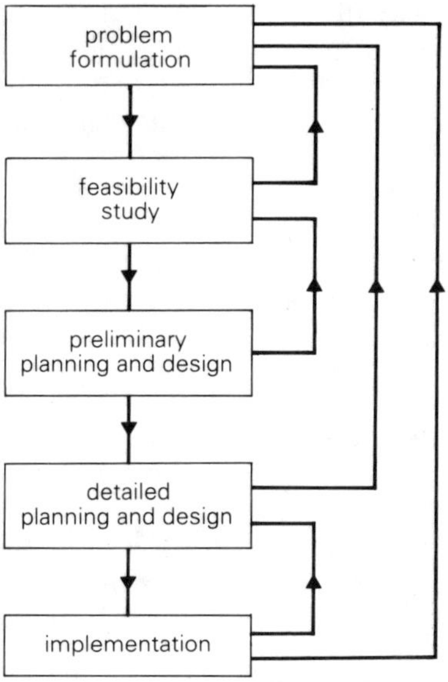

Figure 2.3 Engineering methodology.

Iterative nature of planning and design work

It must be emphasized that planning and design work can rarely, if ever, be carried out by working in a simple sequence through the various steps shown in Figure 2.2. As work proceeds on one of the steps, it may become clear that modifications or improvements are needed in an earlier step. Work may thus have to be suspended on the current step to allow the required improvements to be made. For example, in the preliminary planning and design step it may well be found that the initial problem formulation is too vague to allow useful cost comparisons to be made for the approaches which have been identified in the feasibility study. The entire process shown in Figure 2.3 is iterative, and it is quite normal for the work to proceed by repeated cycling through the steps. In the flowchart, the main feedback loops are introduced as a means of indicating the iterative nature of the process.

Alternative approaches to planning and design

This methodology for planning and design is based on a simple, commonsense approach to solving open-ended problems and has been formalized into the sequence of steps shown in Figure 2.3. The description of the planning and design process in terms of these specific steps is partly a matter of convenience. It would also be possible to group or subdivide the steps of Figure 2.3 to achieve a slightly different description of the process.

For example, the problem formulation step could reasonably be incorporated in the feasibility study. Indeed, it may be advantageous sometimes to proceed with the problem formulation, and the search for feasible solutions, in a more or less iterative manner. More creative results might sometimes be achieved in this way. Nevertheless, a logical separation of problem from possible solutions is important at some stage in the process so that an objective evaluation of alternatives can be made. In the present discussion, problem formulation is treated as a separate step in order to emphasize the importance of critical evaluation in any engineering work.

The steps displayed in Figure 2.3 are usually convenient in practical application and correspond to common engineering practice. Preliminary and detailed planning and design are common steps in most engineering projects, whether large or small in scope, and irrespective of the branches of engineering involved. The need for a feasibility study, at least for large-scale engineering projects, is widely recognized. A formal report on the feasibility study (including details of the problem formulation) is often required by a government department before detailed work on a large-scale engineering project is allowed to proceed.

Although the design process has been described by some authors (Asimow 1962, Ostrofsky 1977, Svensson 1974) in terms slightly different to those used here, the differences are in presentation, rather than in content and intent.

The methodology summarized in Figure 2.3 may be described as a commonsense, rational approach. It may also be criticized because of these very attributes. The convergent, rational process of thinking which is apparently implied by this sequence of steps might also be considered to have an inhibiting effect on creative and innovative thinking. Indeed, numerous examples can be given of significant and successful engineering projects which have commenced *not* with the identification of human needs and a clear definition of objectives, but rather with a good technical idea, which has been followed by a search for possible applications. For example, the search for new applications of materials, devices, and techniques which have been developed in other contexts is a well-recognized and respected engineering activity.

The important point to be emphasised here is that any design procedure is in essence *iterative*. The entry point to the process is not therefore of prime importance. An engineering project may well begin with a good technical idea (i.e. with a feasible solution), and continue with an attempt to identify an appropriate problem. In such a case, the feasibility study is the beginning of the project, as indicated in Figure 2.4.

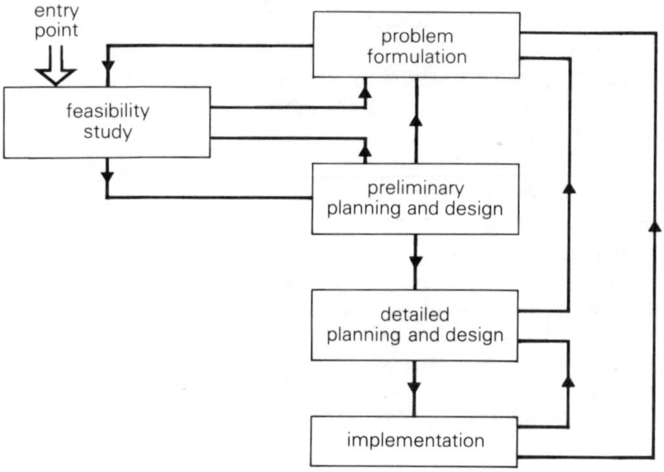

Figure 2.4 'Solution-first' methodology.

Nevertheless, there is one great advantage in adopting the systematic approach indicated in Figure 2.3. Any combination of problem and proposed solution has to be tested properly, by carefully searching for and evaluating a range of alternatives, and also by iteratively adjusting both the problem statement and the proposed solution.

Dealing with creativity

The need for creativity and innovation is particularly strong in the feasibility study, i.e. at the stage when a search is made for new and unusual approaches which are also cost-effective and efficient. However, specific open-ended engineering problems may arise during any step, and deserve creative and innovative solutions. It should be noted that the original problem-solving procedure discussed in Section 2.2 is applicable to such problems.

It follows that creativity is an important attribute for an engineer. Furthermore, the personality of the engineer can have an important influence on the way a project is carried out. The solutions presented by any two engineers to an engineering problem are unlikely to be the same. In the case of a complex problem, they may not even be similar. This is particularly true for planning and design work, which is basic to all branches of engineering. For this reason, engineering work is far more than just the routine use and application of engineering principles and mathematical techniques.

2.4 The problem formulation step

Now that the purpose of the main steps in the planning and design procedure have been discussed, we can go on to consider how each of these steps might be undertaken, beginning in this section with the first step. In Table 2.1 a checklist is given of possible activities which might be undertaken as part of the problem formulation step. These tend to fall into two groups. The first group is concerned with gathering background information. The second group is involved with the actual specification of the problem. We shall consider these activities in turn.

Table 2.1 Problem formulation – checklist.

Identify the problem as part of a wider problem
Identify each engineering system as part of a larger system
Determine the real underlying needs
Gather relevant background information
Search for possible side-effects of the project
Identify constraints
Specify objectives and identify possible conflicts among objectives
Specify any performance requirements and operating conditions
Devise measures of effectiveness

IDENTIFY THE PROBLEM AS PART OF A WIDER PROBLEM

It has already been noted that an engineering project usually begins in response to community needs, and that the objectives are likely to be ill defined. The problem may initially be ill defined because it is stated in very vague terms. However, it may also be poorly defined because it is stated in over-precise terms which imply only one possible solution. As the implied solution is frequently not the only, nor even the best, solution, it is useful to begin by trying to identify the *stated* problem as part of a wider problem.

An example may be instructive. The city of Sydney is separated by a large harbour into a north and a south side. Motor traffic between north and south is catered for by a harbour bridge and a second bridge at Gladesville, a suburb somewhat inland from the city. Calls have frequently been made in the community for a second bridge across the harbour in order to relieve the motor traffic congestion on and around the harbour bridge. What is the real problem? It may not necessarily be *how to construct a second bridge*, but rather *how to improve the cross-harbour traffic flow*. An advantage of this more general view of the problem is that it widens the range of possible solutions. For example, it is now possible to consider a tunnel, a series of smaller peripheral bridges, a widening of the existing bridge, and even various forms of surface (water) and air traffic.

However, it is possible to broaden the problem further: Is the real need for increased traffic capacity between the north and south sides of the harbour? The problem might be formulated as *how to reduce or eliminate the current traffic congestion*. There are various possible alternatives to increasing the effective traffic capacity. For example, it might be possible to reduce the peak usage of the present facilities by means such as staggered starting and finishing times for various industries, the use of flexitime in offices, the imposition of financial penalties for traffic in peak times, and the increased use (by means of various financial inducements) of railways and other forms of public transport.

It is emphasized that the present discussion is not meant to suggest solutions to the traffic flow problems in Sydney. The aim is to emphasize that the generalized problem statement is preferable to the overspecific statement which is likely to be constructed around an implied, pre-chosen solution.

IDENTIFY EACH ENGINEERING SYSTEM AS A COMPONENT OF A LARGER SYSTEM

The process of identifying the problem as part of a wider problem frequently involves identifying an engineering system as part of a larger system. In the example above, the wider statement of the problem has diverted attention from a specific piece of hardware (a bridge) to the larger transport system, of which the bridge would be a component.

It is useful to take the problem-broadening approach much further. In the example of the bridge, is the real problem one of providing an efficient transport system? The very large costs involved in a long-span bridge or tunnel, including associated works such as access roads and traffic rerouting, may be in the order of a thousand million dollars. By alternative use of such resources, with imaginative long-term planning, it might be possible to achieve a more workable city and eliminate the need for heavy traffic between the north and south sides. The discussion now extends into the realm of city planning. We can go further: the use of such quantities of resources to improve transport in an already congested city region might well be queried. If the city transport system were to be improved, the result is likely to be yet more city growth, and in a short period of time a call for further improvements to the again inadequate transport system. An alternative use of the resources could be in decentralization schemes with the aim of relocating industries and population away from the city region and towards regional growth centres.

How far should such arguments be taken? Clearly, we could go even further, beyond the stage of regional planning, and into questions of national and even international planning. At some stage, however, this line of argument becomes unrealistic and breaks down. This occurs when the expanded system is so large that the original problem (in this case, the need for a bridge, or some alternative) ceases to be important or relevant. At that stage, the process of problem-widening has been taken too far.

The problem-widening and system-generalization process is shown schematically for the example of the bridge in Figure 2.5. The limit beyond which the proposed project ceases to be of significance defines an approximate **interest boundary** within which alternative approaches will be found, and within which alternative uses of the resources for the project should be considered. In the example of the bridge, the very large costs suggest that the interest boundary extends somewhat beyond the city system and perhaps even the state. A large bridge is likely to become a national consideration, at least in the case of a relatively small country.

We now turn to a much smaller project, such as the redesign of an ineffective and hazardous traffic intersection in a suburban area, where the interest boundary is correspondingly smaller. The traffic intersection might initially be regarded as part of the local traffic flow, which can in turn be

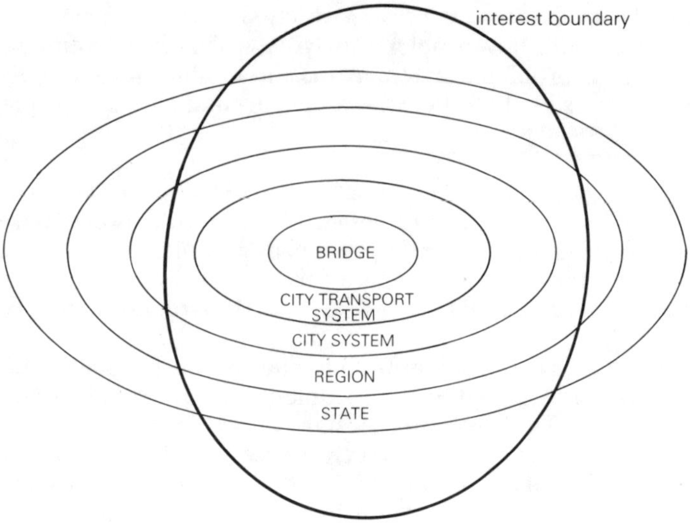

Figure 2.5 Problem-widening: the example of a large bridge.

considered as part of the traffic system in a region of the city. The situation is shown schematically in Figure 2.6. In this case the interest boundary at which the traffic intersection becomes only of marginal interest may encompass the surrounding streets and possibly the regional traffic flow, since a reorganization of the traffic flow could provide the basis for a solution of the problem. On the other hand, conditions at one small intersection are unlikely to be of great importance to the efficient flow of traffic throughout the city.

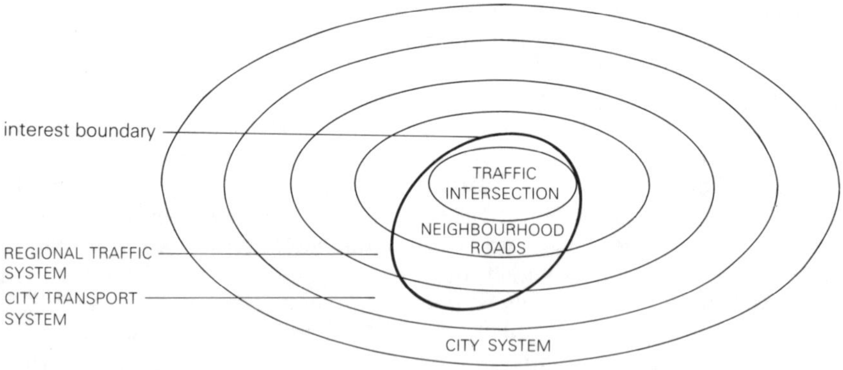

Figure 2.6 Problem-widening: the example of a small traffic intersection.

In essence, the boundary-widening process should be halted when changes to the original system under consideration do not have a significant impact on the system just outside of the boundary.

DETERMINE THE REAL UNDERLYING NEEDS

The real needs are not always those which are initially perceived. The process of problem-widening and the identification of the system as a component in a wider system often lead to a better understanding of the real needs. In the bridge example, the need is not really for a bridge *per se*, but for a reduction in traffic congestion or, possibly, a more efficiently functioning city. If the real needs can be clarified, the problem can be formulated in a much more useful manner.

In some situations it is useful to study the underlying needs directly. As an example, consider the design of a backyard swimming pool, undertaken by a consulting structural engineer for a private client. Before the size and shape and other details of the pool are chosen, the real needs of the client have to be clarified. The pool may be needed for practice for competition swimming, for high diving, or for social entertaining and relaxation. It might also be required purely for prestige reasons. In each case, a different design would be required.

The attempt to identify underlying needs sometimes moves attention away from purely technical problems towards social problems and issues, especially when the proposed engineering project is large in magnitude and cost. This is of course inevitable when large sums of public money are to be spent. As the problem-widening and system-widening procedure is undertaken, the questions which arise may thus acquire an increasingly political and social flavour.

GATHER RELEVANT BACKGROUND INFORMATION

When a team of engineers and other professionals commences work on a new project it will often be found that they lack relevant background information. It is essential that such information be gathered as early as possible, and before the problem is formulated. The information may be scientific and technological, but it may also be non-technical and even sociological or political in nature.

The data-gathering exercise may require use of libraries, textbooks, encyclopedias and databases. Sometimes it will be necessary to gather field data or conduct laboratory tests, as for example when the properties of a foundation material are investigated by a soils engineer.

The type of information an engineer may need to gather is almost unlimited in scope. For example, in the design of a replaceable turf wicket for use in first-class cricket matches, the necessary background information would include details of the conduct of the game of cricket and the laws governing play. Also needed would be detailed horticultural information concerning the cultivation of appropriate grass, the depth of soil needed, the type of soil, and the compacting and wearing properties of turf. In unusual projects where the engineer does not have adequate background knowledge, it may become necessary to work in association with consultants who are expert in the relevant field.

Considering as a further example the relocation of access roads in a suburban area, we may see the need for background information concerning the soil properties along the proposed routes, as well as the present and projected traffic volumes. To obtain this information a programme of soil testing and traffic counting may have to be undertaken. However, equally relevant background information might concern the social structure of the region and where the proposed routes lie in relation to natural boundaries between communities. The political implications of the project, measured in terms of votes gained and lost, may also be very relevant to the decision to proceed with the project.

SEARCH FOR POSSIBLE SIDE-EFFECTS

All engineering projects are designed to bring about some changes in the world in which we live. Although the intended aim is to improve the infrastructure in order to satisfy community and individual needs, it is inevitable that there will be some side-effects. These may be subtle or obvious, desirable or undesirable. In the case of a large-scale engineering project, a range of desirable and undesirable side-effects can be expected to result simultaneously.

It is important that all relevant side-effects be identified and taken into account in the evaluation of the costs and effectiveness of any project.

Some disturbance of the environment is inevitable whenever a large-scale project is undertaken, and environmental side-effects have received much attention in recent times. When a road or freeway is constructed, either in an urban or rural environment, some destruction of the surroundings is involved. This is likely to involve the demolition of buildings in the case of a city road. It may involve cutting into a hillside or filling in part of a valley floor in the countryside. Even when a large-span bridge or viaduct is used to lessen the impact on the environment, there is likely to be a strong visual impact. Of course, this impact may prove to be very pleasing.

Technical side-effects may be advantageous or disadvantageous. For example, the provision of lift wells and stair wells in a multistorey building can improve the resistance of the structure to lateral wind load. On the other hand, the increased stiffness may result in increased inertial forces if the building is subjected to earthquake.

Non-technical side-effects which may be significant and need to be considered include increased or decreased job opportunities. These may be either permanent or temporary. Factors as diverse as the creation of a tourist industry following the construction of a monumental project, or the creation of water sport and recreational facilities in connection with the construction of a dam, are examples of non-technical effects which need to be identified in the initial problem formulation step.

IDENTIFY CONSTRAINTS

Constraints are limitations or minimum performance requirements which have to be satisfied when the project is carried out. Constraints may arise for a variety of reasons such as the limited availability of certain resources, the unacceptability of certain side-effects such as atmospheric pollution, or technological limitations. Various constraints may be applied to the design of a system component so that it can fit in with, and work in harmony with, other components in a system. Other non-technical constraints may arise from legal requirements or industry standards.

Examples of legal constraints are the emission control requirements for automobile engines, the minimum fire protection requirements in office buildings, and the noise and thermal controls which have to be provided according to building regulations in apartment houses. Height limitations on trucks which arise from the clearance to existing bridges are examples of physical constraints. Political constraints are not so easy to document, but clearly apply in, say, the siting of freeways and airports.

DEFINE OBJECTIVES AND IDENTIFY CONFLICTS AMONG OBJECTIVES

It follows from the previous discussions that the objectives, or purposes, of an engineering project should only be defined after considerable investigation has been carried out. In particular, the process of identifying relevant systems as components in a wider system, and the related activity of identifying the problem as part of a wider problem should already have been undertaken. It will also be clear that the objectives of an engineering project should not be specified in terms which are either too specific or too general.

Most engineering projects are multi-purpose, and it is frequently the case that the various objectives conflict one with another. For example, a dam may be designed for the joint purposes of providing water storage, producing hydro-electricity, and reducing or preventing downstream flooding. In planning the operating policy for the dam it will be found that these purposes conflict with each other. For effective storage, the dam needs to be kept nearly full of water, while to serve the purpose of flood prevention the dam needs to be in a nearly empty state to cater for unexpectedly severe rainfall. In any engineering planning and design work, the twin requirements of maximum performance and minimum cost are in themselves conflicting objectives. It is important that all such conflicts among the objectives be recognized and allowed for.

SPECIFY ANY PERFORMANCE REQUIREMENTS AND OPERATING CONDITIONS

In the case where the project involves the design and production of a system or device, such as a bridge or a transmission unit for an automobile, it is necessary to determine both the conditions under which it will be required to operate and the level of performance which will be expected.

If, for example, a pipeline is required to move natural gas from a production field to a consumption site, the performance requirements for the pipeline would involve the total quantities of gas to be moved, the average and peak rates of supply, and the expected life of the operation. This information would be used in the planning and design of storage tanks, pumps, and pipeline.

The operating conditions of interest might be the ambient temperatures throughout the year, the magnitude of any likely earth subsidence over the possible routes, the likelihood of earth tremor, and possibly the chemical properties of the gas. This information would be taken into account in the design of the pipeline system.

Safety and reliability are important performance requirements which always require very careful consideration. It is now recognized that absolute safety is not an achievable or feasible objective in any field of engineering. Even with the introduction of ingenious, fail-safe concepts, the probability of failure of any real engineering system cannot be reduced to zero. Furthermore, the incremental cost of reducing the risk level increases sharply as the risk level decreases; eventually a stage is reached where the cost of achieving further improvements in safety become impossibly high. Engineering design thus involves, in a very real sense, calculated risk, and, in rare cases, the prospect of malfunction and failure.

In the original design of the Tasman bridge in Hobart, careful consideration was given to the possibility (which was considered very remote) of a ship collision with the main piers. Indeed, steps were taken in the design of the superstructure to limit the effect of such a collision (New *et al.* 1967). Nevertheless, the unlikely event occurred in January 1975, when a ship collided with the bridge, with catastrophic and tragic consequences.

Given that an element of risk is inevitable in engineering work, one of the difficult tasks in planning and design is to choose the levels of risk which can be considered acceptable. If the safety requirements are set too low, the rates of occurrence of malfunction, unserviceability and failure will be unacceptably high. On the other hand, if the safety requirements are set too high, the cost of achieving the implied performance levels can become impossibly high and the project will cease to be financially feasible.

The risk levels adopted in engineering work usually have an indirect effect on the entire community, in regard both to the safety and the cost of the common engineering infrastructure. It is for this reason that national standards give guidelines and set minimum safety and performance requirements in many fields of engineering.

DEVISE MEASURES OF EFFECTIVENESS

Measures of effectiveness are used in both the feasibility study and in the preliminary design as a means of evaluating and ranking alternative competing approaches or designs or plans. A measure of effectiveness can thus be developed for each specific objective and then applied to the alternative approaches in turn to determine which one best meets that objective.

In some projects, total financial cost may be an adequate measure of effectiveness. This might be the case when minimum performance requirements and operating conditions have been clearly defined, as for example in the design of an electronic component which will be mass-produced. In other situations the measures of effectiveness may have to take account of matters as diverse as aesthetics, environmental effects, and risk of injury or loss of life.

Even when at first sight financial cost appears to be the appropriate measure of effectiveness, conditions can become quite complex on closer analysis. For example, in the design of a bridge superstructure total cost might be considered the best measure of effectiveness. It would then appear to be a simple matter to choose from among the alternatives of, say, a steel truss, a steel plate-girder, a reinforced concrete (*in situ*) girder, and a precast prestressed concrete girder. Purely in terms of cost of construction, the plate-web girder might be the cheapest solution. However, when maintenance costs are also taken into account for the expected lifetime of the bridge (possibly 50–100 years) the reinforced concrete solution might prove to be cheapest. But other factors usually have to be considered. Minimum disruption of traffic during construction might be an important objective in a city construction site, and a prefabricated truss solution might then be favoured because of minimum construction time. Then again, if importance is also given to appearance, as well as cost and time of construction, the precast prestressed concrete solution may be preferred.

The manner in which the measure of effectiveness is formulated, and in particular the parameters which are included, are enormously important in determining the path which an engineering project will follow.

In some circumstances engineering planning and design work is carried out without the explicit use of a measure of effectiveness, the choice of 'best' design, for example, being made subjectively by the design engineer. The justification given for such an approach is that it is too difficult to obtain satisfactory quantitative measures of effectiveness. In other cases the measure of effectiveness is tacitly taken to be monetary cost. However, errors of judgement can very easily be made unless an attempt is made to prescribe a measure of effectiveness. The attempt to define appropriate measures of effectiveness is always valuable, and gives considerable insight into the project, even when a purely quantitative criterion is not found.

COMPLETION OF THE PROBLEM FORMULATION STEP

It has previously been emphasized that planning and design and engineering problem-solving must be undertaken in an iterative manner. It will usually be found that the various activities involved in the problem formulation step can also be best undertaken iteratively. Thus, the type of background information which will be needed will often become clearer after some attempt has been made to identify side-effects and problem constraints. Likewise, some of the constraints and side-effects may only be identified after some background information has been gathered.

At the end of the problem formulation step, background information on the proposed project will have been gathered, the problem will have been put into an appropriate context, and specific details of required operating conditions will have been established, together with the required levels of performance. The choice of suitable measures of performance will also have been considered.

As already noted, it may well become necessary to return to the problem formulation at a later stage when, for example, inadequacies are detected during the feasibility study or the preliminary planning and design.

2.5 Feasibility study

The purpose of the feasibility study is to determine whether or not the project can be carried out successfully, and if so to provide a short list of promising alternative approaches or 'concepts' which can form the basis of the project work. The procedure in the feasibility study is to search out a range of alternative and preferably contrasting approaches, and then to investigate each in turn to determine which are feasible, i.e. would result in satisfactory performance under the specified operating conditions, without violating any of the constraints. The alternative feasible approaches can also be compared with each other with regard to cost and effectiveness. A checklist of steps which may be appropriate to the feasibility study is given in Table 2.2, and are discussed below.

Table 2.2 Feasibility study – checklist.

Check available resources
Investigate and quantify the constraints
Develop as many promising concepts as possible
Compare alternative concepts by means of the measures of effectiveness and progressively eliminate the non-competitive approaches
Modify problem formulation as necessary
Scrap or defer project if no feasible approaches can be found

CHECK AVAILABLE RESOURCES

The project can only be implemented if there are sufficient resources. The feasibility study should therefore include a careful survey of available resources. As the type of resources and their availability can well affect the approaches which will be developed, it is desirable for the resource survey to be carried out at the beginning of the feasibility study. The resources which need to be considered include both financial and human ones. Engineering and scientific expertise are resources, as are the relevant trade and craft skills, and any special materials needed for construction or manufacturing purposes. Time is also a resource to be considered, as engineering work always has to be completed in limited time.

INVESTIGATE AND QUANTIFY THE CONSTRAINTS

The technical and non-technical constraints which were identified in the problem formulation step need to be investigated and, where appropriate, quantified. Many of the constraints which apply to engineering design work are legal–technical, and are quantified in codes of design practice and in legislation. Thus, in the design of a city building, details concerning overall height limitations, minimum water and sewerage requirements, vertical transport, and acoustic and thermal insulation, are usually dealt with in relevant ordinances and building acts, while restrictions concerning fire resistance, allowable deformations, and structural safety may be covered in the structural design standards and appropriate codes of practice.

Other constraints arise naturally. For example when a plant for the manufacture of cement is set up, the manufacturing process is constrained by the quality of the local raw materials (clay and limestone).

DEVELOP AS MANY PROMISING CONCEPTS AS POSSIBLE

This is probably the key step in the entire process of planning and design. The success of the project depends on the quality of the solutions that are proposed in this step. It is highly desirable for a diverse and varied set of possible approaches to be developed. Innovative and creative new approaches, as well as previously used and proven approaches, should be included. It is by no means certain whether the innovative approach will prove to be superior to the well-tried standard approach; both types have to be investigated.

The technique of problem-widening, already discussed in regard to the problem formulation step, should prevent over-specification of the problem, and so leave open the possibility of using new and unusual approaches. In Chapter 4 some techniques are described for developing a range of concepts and approaches for this stage of the process.

PROGRESSIVE ELIMINATION OF NON-COMPETITIVE APPROACHES

A preliminary screening of the alternative concepts is made in order to eliminate any which would not lead to feasible, economic solutions to the problem as already formulated. Analysis is used to check that the performance will be satisfactory under the required operating conditions, i.e. as a check that an approach is feasible. For example, in the design of a highway bridge the normal operating conditions will be the day-to-day service loads which are imposed by normal traffic. The performance requirements might include limits on deflections and vibrations produced during normal traffic flow, as well as the ability to carry, without failure, overloads up to twice the magnitude of the normal traffic loads. Techniques of structural analysis are used to calculate the magnitude of deflections and vibrations, and the load capacity of the structure, to check that these fall within the required limits.

A further requirement for feasibility is that all the constraints be satisfied. An example of a constraint on the bridge design might be the restriction that no piers be placed in the main channel region used by shipping.

MAKE MODIFICATIONS AS NECESSARY TO THE PROBLEM FORMULATION

The need for an iterative approach to problem-solving has already been emphasized. In particular, the possibility of modifying the problem formulation should be considered during the feasibility study. As work goes on and more approaches are investigated, the understanding of the engineer is inevitably improved. With improved understanding it is possible to achieve a better problem statement. In particular, the measures of effectiveness can often be made more apt and more appropriate to the overall project.

SCRAP OR DEFER PROJECT IF NECESSARY

If none of the approaches is feasible, then the project may have to be scrapped unless further alternatives can be found. Another possibility is for the problem to be reformulated with a more modest, or more realistic, set of goals.

2.6 Analysis and design

An analysis of the behaviour of a candidate system design is used to check whether the performance will meet the specified requirements. The check may well lead to an adjustment of the design details, followed by a further check for feasibility. The place of analysis in this design process is shown in Figure 2.7. Trial design details for the system are chosen, often on the basis of design experience, but also using any specialized techniques which may apply in the relevant fields of engineering. An analysis is then carried out to determine whether the performance requirements are satisfied. If not, a trial adjustment has to be made to the design details and a new analysis carried out in the search for a feasible solution. Analysis plays a comparable role in a planning problem.

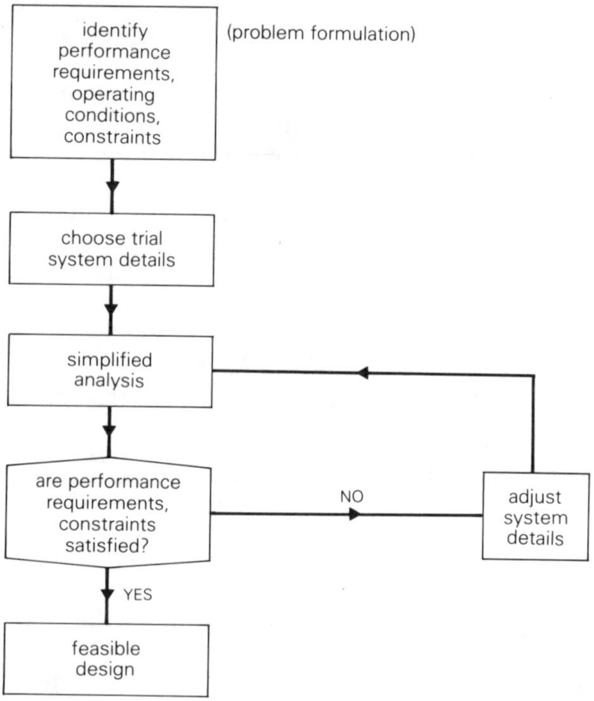

Figure 2.7 The role of simplified analysis in the feasibility study and preliminary design.

As this work is in the preliminary feasibility study step, simple and approximate methods of analysis need to be used at this stage. As already noted, a range of analytic techniques is ideally required for use in the different stages of planning and design.

2.7 Preliminary planning and design

The preliminary planning and design step is essentially a continuation of the feasibility study. It brings the search for the most appropriate approach to a conclusion. Each of the alternative approaches shortlisted in the feasibility study is investigated in turn and details are worked out to the extent necessary to enable them to be compared and ranked by means of the measures of effectiveness. In order to obtain an accurate cost estimate and a detailed estimate of the effectiveness of a possible approach, a great deal of planning and design work needs to be undertaken, which is costly. If the engineer goes into too much detail in the preliminary planning and design, in order to compare and rank the alternative approaches, this step becomes unnecessarily expensive and time consuming. On the other hand, if alternatives are eliminated on insufficient information, the most appropriate alternative may not be chosen.

A progressive approach is usually appropriate, whereby the concepts are first developed to an initial stage which allows the least competitive alternatives to be eliminated. The development is then carried to a more detailed stage, so that further alternatives can be eliminated. In this way, progressively fewer alternatives are developed to an increasing level of detail, until the final choice is made.

Although attention is focused on the original alternative concepts which arose from the feasibility study, the search for new and better alternative approaches should not be discontinued. As work goes on, the build-up in expertise and background knowledge can lead to the development of different and better alternatives. These may be completely new or modifications to the original approaches. They should be brought into the list of alternatives, if possible as replacements for less viable ones. Increased expertise and additional background information may also lead to improvements and changes to the original problem formulation and the measures of effectiveness.

2.8 Detailed planning and design

In the detailed planning and design step, full details have to be worked out to allow implementation of the project. For example, at the end of the preliminary design of a reinforced concrete bridge the form of construction and the overall dimensions will have been chosen. In the detailed design, full information on each group of reinforcing bars, on steel fitments, finishes, surfaces, handrails, bearing pads, stormwater pipes, etc. have to be worked out so that complete construction plans and specifications can be prepared.

The process of detailed planning and design is also iterative. It may require the choice of trial details, followed by detailed and often complex engineering analysis. The trial procedure shown in Figure 2.7 needs to be followed, however using accurate methods of analysis and supplemented by a check to ensure that the proposed solution is the best that can be found. Iterative adjustment is made to the design details to produce either a progressive decrease in costs or an improvement in performance.

The modified procedure for detailed design is shown in Figure 2.8.

Adjustments are now made progressively to the design details in order to achieve an improvement in performance, and new analyses are then carried out. This process is continued until an effective and economical design is achieved, for which the performance meets the design requirements.

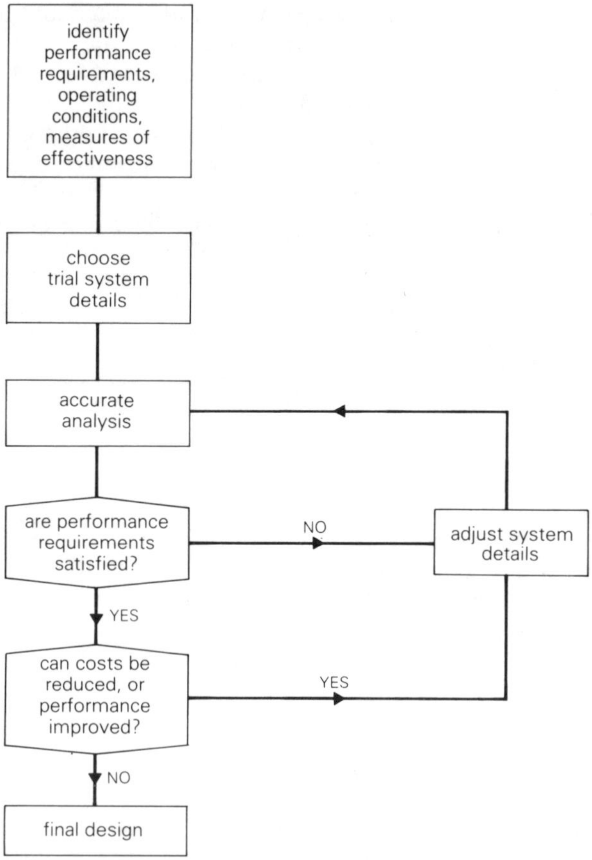

Figure 2.8 Detailed design.

It is in the detailed design step that certain design optimization techniques sometimes prove to be useful. If the behaviour of the component lends itself to accurate theoretical modelling, it may be possible to improve the design by mathematically optimizing the parameters of the component. The process of optimization is discussed briefly in relation to analysis in Chapter 3, while Chapters 7 and 8 deal with various optimization techniques.

In the manufacture of a component, it may be advisable, depending on the type of component, its potential cost, and the number to be produced, to construct prototypes and test and modify them as an adjunct, or alternative, to the theoretical analyses.

At all stages of the detailed design it is necessary to check that the design constraints are not violated.

Even during the detailed planning and design phase it may be possible to make improvements to the decisions made in the problem formulation phase.

In some situations severe constraints may add disproportionately to the cost or detract greatly from the effectiveness of a component. A return to the problem formulation phase is then necessary. For example, in the design of a concrete arch dam, good rock foundations are required in the sides of the valley to carry the enormous thrusts due to hydrostatic pressure. In the detailed design phase it may become clear that the rock has insufficient strength, thus making the arch dam concept unworkable. Alternative design concepts then must be considered, such as a mass concrete dam wall.

The scope of the detailed design work on each component should always be broadened to consider the overall operation of the parent system. If the constraints imposed on the design of the component have a disproportionate effect on the cost and effectiveness of the overall system, then some relaxation of those constraints has to be considered, together with corresponding adjustments to the design of companion components.

The important final step in the detailed planning and design phase is full documentation of the plans and designs, with a permanent record of relevant calculations and analyses and any other investigations which have been used to produce the final details of the project.

2.9 Implementation

Implementation is the second main phase of an engineering project, and has been treated here as the final step in the engineering methodology. Details of the implementation step depend on the nature of the project. Implementation may require the mass production of a device or the one-off construction of a large and complex structure, or possibly the initiation of a new production process. Special engineering management techniques will usually be required. A number of branches of engineering are involved with project implementation, such as industrial engineering, construction engineering, etc.

2.10 Example: planning and design for a city water supply system

To illustrate the important initial steps of problem formulation and feasibility study in the methodology for planning and design, we consider briefly an engineering project concerned with ensuring an adequate water supply for a small coastal city in a very dry climate. We choose the city of Adelaide in South Australia as a focus for the discussion, in order to give some realism to the example. However, it must be emphasized that our considerations are far too superficial to lead to valid conclusions regarding this city.

With a city population of the order of 1 000 000, the current water supply system is capable of supplying around 200 000 ML per year with reasonable security. The initial statement of purpose is to increase this capability to 330 000 ML by the year AD 2020. This target figure has been obtained by extrapolation from current data on water use.

Problem formulation

When the need for an engineering project is initially perceived in the community, the statement of purpose is likely to be vague and ill defined. It may alternatively be misleading by being too specific. In either case it is useful to distinguish between the **apparent problem** and the **real problem**. In order to identify the real problem we consider those activities listed in Table 2.1 which are relevant to the present problem.

IDENTIFY THE PROBLEM AS PART OF A WIDER PROBLEM

It is important to recognize that the supply of water to the city of Adelaide is part of the larger problem of providing for the water needs of the entire state of South Australia. Solution of Adelaide's water problems without reference to the water needs in other settled areas could well lead to non-essential use (i.e. waste) of water in the city, while the requirements for water in country areas (for irrigation) and in other city areas (Port Augusta and Port Pirie) are not met.

It may be important to identify the problem as part of an even wider problem: the regional management of water resources throughout the Murray–Darling river system which extends through three eastern states on the Australian continent.

Similarly, the engineering system (the city water supply system) might be considered as a component in the wider system (the water resources of the state, or of the Murray–Darling system). The water supply system might also be considered as a component in the city system, together with a variety of other components such as the sewerage system, the industrial regions, and the residential suburbs. This line of discussion would highlight the way water is used in the city, a matter which will be considered below.

DETERMINE THE REAL UNDERLYING NEEDS

To formulate our problem adequately, we need to consider the need for water in the city, and possible changes in use over the next thirty or so years. Water needs, and use, fall into a number of categories, some examples of which follow:

- **domestic:** drinking, cooking, washing, cleaning, flushing of toilets, watering of domestic lawns, use in private swimming pools
- **industrial:** cooling and heating operations, chemical processes, cleaning
- **public use:** street cleaning, watering of public lawns and gardens, public swimming pools
- **other:** irrigation for market gardens and nurseries, firefighting

Although it is relatively easy to identify such uses, it is much more difficult to evaluate the extent of the needs. When is a need a necessity, and when is it a desire or even a luxury? A surprisingly large proportion of the water for domestic consumption in Adelaide is used to grow exotic European lawns, trees, and flowers in a very dry climate. This raises the important question of what is a legitimate need.

GATHER RELEVANT BACKGROUND INFORMATION

The present size and distribution of the population, the location of industrial regions, the current pattern of water use (both spatial and temporal) for domestic, industrial, and other types of consumption are examples of relevant background information. Expected demographic changes over the planning period are clearly very important, as are predicted changes in the nature and extent of industry.

The present sources of water supply, the rainfall and runoff patterns in the city and throughout the existing catchment areas need to be investigated, as well as potential new surface and underground and remote sources of supply.

IDENTIFY POSSIBLE SIDE-EFFECTS

The side-effects which may arise from this project will depend on the approach or approaches which are finally adopted. The environmental effects of dam construction are well known. The effects following on from depletion of underground water by aquifer pumping may not be quite so immediately obvious but may include land subsidence, and dewatering of bores which use the same aquifer. Over-use of river water may well lead to reduced flows, silting, environmental effects on the biota, or even cessation of flow and closure of the river mouth in summer months.

A variety of further effects to be considered include increases in salinity levels in the rivers, the need for more complex and more expensive treatment procedures to achieve adequate water quality, and the inhibiting effect on industrial expansion of an inadequate quantity or quality of water.

IDENTIFY CONSTRAINTS

Some of the constraints to be considered include state and federal regulations regarding water supply and land resumption, World Health Organization standards for water quality, health regulations, city planning and zoning laws, inter-state agreements concerning the use of water from the Murray River, and environmental impact legislation.

SPECIFY OBJECTIVES

The objectives of the project are not likely to be stated in the original terms of providing 330 000 ML of water per year to the city of Adelaide. A more general, improved statement might be:

> *to match, as closely as is economically feasible, the quantity of water, of appropriate quality, to the realistic and changing needs for water in the community over the planning period of 30 or more years.*

In this formulation, which is much less specific than the original, both the supply and the demand are allowed to be variable quantities.

The prediction of realistic demand figures now becomes a task to be undertaken as part of the project.

The concepts of economic and technological feasibility are linked together, and are implied in the statement of objectives. Thus, it is not feasible, and it is not therefore stated as an objective or as a design constraint, that water must always be available without restrictions, even in times of severe drought. A reliable minimum supply is nevertheless an important objective.

SPECIFY PERFORMANCE REQUIREMENTS AND OPERATING CONDITIONS

The water supply system has to be developed in a manner which allows it to operate adequately at all times in the planning period (up to AD 2020) under a wide range of conditions, including extended periods of drought and periods of flood in the upper reaches of the river system, and possibly even the simultaneous combination of flood and drought in different parts of the upper reaches. For this wide range of conditions it will be necessary to decide on a minimum quantity of water to be supplied on a daily and seasonal basis to the city. The quality of the water supplied also needs to be specified, together with other secondary matters such as minimum water pressure.

DEVISE MEASURES OF EFFECTIVENESS

Although total cost, expressed in current monetary value, will be an important measure of effectiveness, there will be other aspects which need to be taken into account when alternative approaches, and combined approaches, are evaluated. For example, reliability, measured as the expected number of days of restrictions per year, and adaptability, i.e. the ease with which the system can be changed in response to unexpected developments, are important additional measures. Adverse environmental effects also need to be considered, if these are markedly different for alternative approaches.

A combined measure of effectiveness could well be monetary cost per unit of water supplied, with penalty costs added to allow for non-monetary effects such as limited reliability, environmental effects, etc.

Feasibility study

ESTIMATE AVAILABLE RESOURCES

In this example, the resources which need to be considered include the financial and technical resources which will be needed to carry out the project, but they also include the water resources which are potentially available in as yet unused catchment areas, and the quantity of water still potentially available from the Murray–Darling system.

INVESTIGATE AND EVALUATE CONSTRAINTS

Some of the constraints which might apply to the problem have already been listed above. It is necessary to deal specifically with them and to quantify them, in order to determine how they will affect feasible solutions.

DEVELOP PROMISING CONCEPTS
An important aspect of the feasibility study obviously involves the development of various standard approaches for increasing the quantity of water available. This would involve the investigation of currently unused catchment areas, possible sites for dams, and the possibility of pumping water from underground sources. We shall not discuss these approaches in detail because this would involve much unnecessary technical detail, coupled with a detailed evaluation of very specific local conditions.

Unconventional approaches might include bringing in icebergs from the southern oceans, re-use of sewage, desalination of sea water, and the installation of domestic rainwater tanks. It is also of importance to consider less obvious approaches such as *not* attempting to meet fully the projected increased demand. An alternative approach would be to try to inhibit the rate of increase in demand for water, for example by informing the community of the real costs of excessive water demand, education programmes on reduced use of water, and a policy of increasing water price.

COMPARE ALTERNATIVE APPROACHES
In the case of complex engineering projects such as the one under discussion, a mixed strategy will often be appropriate with a whole range of approaches being adopted for simultaneous implementation. The comparison of alternatives in this case should lead not only to the elimination of non-competitive approaches, but also to the study of various compatible groups of approaches.

References

Asimow, M. 1962. *Introducton to design*. Englewood Cliffs, NJ: Prentice-Hall.

Biswas, A. K. 1976. *Systems approach to water management*. Tokyo: McGraw-Hill.

Blockley, D. I. 1980. *The nature of structural design and safety*. Chichester: Ellis Horwood.

Hall, P. 1980. *Great planning disasters*. London: Weidenfeld and Nicolson.

Jones, J. C. 1981. *Design methods*. New York: John Wiley & Sons.

McLoughlin, J. B. 1969. *Urban and regional planning, a systems approach*. London: Faber & Faber.

New, D. H., J. R. Lowe & J. Read 1967. The superstructure of the Tasman Bridge, Hobart. *The Structural Engineer* **45**, 81–90.

Ostrofsky, B. 1977. *Design, planning, and development methodology*. Englewood Cliffs, NJ: Prentice-Hall.

Svensson, N. L. 1974. *Introduction to engineering design*. Randwick: University of New South Wales Press.

Problems

2.1(a) Identify two major bottlenecks in road traffic in the city where you live (or in a city with which you are familiar). Determine on which days and at what times these bottlenecks do not cope adequately with the traffic. (What does ‘adequately’ mean in this context?)

(b) Make a list of the kinds of background information you would want to gather in order to make a problem statement concerning these traffic bottlenecks.
(c) Make a tentative statement of the problem(s).
(d) List six different approaches for solving one of these bottleneck problems.
(e) What might be used as measures of effectiveness in comparing the approaches?

2.2 A level crossing occurs where a public roadway crosses a railway line without grade separation. Details of level crossings vary from country to country; however, they are always very dangerous points on the road system and are responsible for serious accidents.

Carry out a feasibility study, aimed at reducing the accident rate at level crossings by 50%. Use appropriate local details in your study.

2.3 It is 'logical' (and 'traditional') to construct the frame of a multistorey building from the foundations upwards, floor by floor.

Is this the only way?
Is it the best way?

2.4 You have been engaged as a consultant to design a nuclear fallout shelter in the back yard of a private residence. Work through the steps involved in the problem formulation and feasibility design phases and come up with a preliminary design concept. Identify the data that you need in the process.

3 Basic system concepts

The systems approach provides an effective framework within which the engineering work of planning, design, and management can be conducted. This chapter introduces a number of basic system concepts.

3.1 Dealing with complexity

Most engineering projects are far too complex for the associated planning and design work to be undertaken as a single operation. It is therefore necessary to break this work down into a number of interrelated tasks and sub-tasks which can be handled separately by individuals or small groups of engineers and possibly other professionals working independently.

In this way, a large project which requires many thousands of man-hours of design, planning and analysis work can be completed by a large team in a very short time. Even in the case of a relatively small project, it is advantageous to organize the work so that the processes of planning and design can be carried out as a number of separate tasks.

However, great care must be exercised in the initial breakdown of engineering work in order that the parts will eventually come together smoothly to produce the desired result. Careful coordination and management of the work are required if this end is to be achieved.

As the scope and complexity of the work increases, so too does the need for systematic project planning. A systems approach to project planning and to large-scale design has gradually evolved in a variety of engineering disciplines. Impetus was given to this development by the need to coordinate very large projects such as the NASA space programme in the United States. However, significant advantages are also achieved when the systems approach is applied to relatively small projects.

In the following discussion, a number of concepts are introduced which are basic to the systems approach to planning, design and management. It should be emphasized that the use of these systems concepts is not restricted to engineering. They find frequent use in fields as diverse as biology, corporate planning, and business management.

3.2 Systems and processes

The word **system** has already been used extensively in previous chapters of this book in an informal way in phrases such as **water supply system** and

communications system. In everyday language, 'system' is a vague word, applied to any complex thing which is organized in some way, but which is difficult to specify precisely. In technical discussions too, the word tends to be used when a more precise, specific term is not readily available.

In the present discussions, the term will be applied to any collection of interrelated and interacting components which work together in an organized manner to fulfil a specific purpose or function. For example, the water supply system already referred to consists of many components such as dams, reservoirs, pumping stations, networks of pipes, and filtration and treatment works. The system functions to provide adequate quantities of good-quality water to the various consumers in the community.

The word **process** is also used frequently but vaguely in both technical and non-technical discussions. In terms such as **manufacturing process** and **chemical process** the word refers to a progressive change with time, which may take place in a piece of matter or in a complex system. In this text, a process is taken to be an activity which is executed progressively in time. For example, the manufacture of an automobile is a process. The construction of a building or a bridge is likewise a process, as is the production of electrical energy in a power station.

Engineering systems are often designed for the specific purpose of executing a process. For example, a chemical processing plant is designed to bring about a sequence of chemical reactions, i.e. a chemical process, which transforms a number of raw materials into desirable end products. A great deal of engineering work is concerned with processes which arise from the operation and management of engineering systems. The interest is often as much in the process which is executed by a system, as in the system itself.

System design and process planning

Although there is no widely accepted convention regarding the usage of the terms **planning** and **design**, the latter term tends to be associated with systems, and the former with processes. Here, we shall generally refer to **process planning**, and to **system design**, which conforms to the terminology introduced in Chapter 1.

System design has already been discussed in some detail in Chapter 2. In particular, the important (but subsidiary) role of system analysis in the design process has been pointed out in relation to Figure 2.8.

Many of the concepts which are applied to systems have equivalents which relate to processes. However, we shall concentrate attention on systems concepts in the following discussion for the sake of brevity. The parallels, as they relate to processes, will in most cases be obvious.

3.3 Systems

Sub-systems, components and system hierarchy

It is characteristic of an engineering system that its components are themselves made up of a variety of sub-components, so that the components may also be regarded as systems. For example, the water treatment plant in the water supply system is composed of a variety of components such as mixing and holding tanks, control valves, measuring devices, etc. The term **sub-system** is also used to refer to a component.

It is also characteristic of an engineering system that it will itself act as a component in a larger system in conjunction with other components. Thus the water supply system, together with various other engineering systems, such as a transportation system, a communications system, and a sewerage system, are components of the city system.

The term **system hierarchy** describes this typical structure whereby a higher-level system is composed of components which are themselves systems composed of sub-systems, and so on. The concept of system hierarchy was used informally in Chapter 2, where, as a step in the problem formulation procedure, we identified an apparent engineering problem as part of a larger problem, and the engineering system as a component of a larger system.

System boundaries

In many engineering applications the various components of a system are clearly separated from each other by distinct physical boundaries. The components of the water supply system are clearly distinguishable from each other. However, the boundaries between components are not always so clear cut. In some situations, arbitrary boundaries may have to be introduced by the analyst in order to break a complex system into a number of smaller, manageable components. For example, arbitrary boundaries are introduced in the analysis of massive structural systems. Thus, in the finite-element analysis of the combined effects of water pressure and earthquake motion on an arch dam, the wall is conceptually broken up by a systematically located set of surfaces into a very large number of small block-like pieces, or 'elements'. By analysing first the behaviour of the individual elements, and then the way they interact with each other in combination, the behaviour of the complete dam can be determined. The mathematical breakdown of the system into component elements in this case does not relate to any physical boundaries and may be quite arbitrary from a physical point of view.

As a further example, the keel and hull can usually be identified as separate components in a ship; however, an exact boundary between these components may in some cases be difficult or even impossible to define on any physical basis. An arbitrary boundary surface may have to be chosen if a detailed structural analysis is to be carried out to investigate the in-service performance of a proposed new design.

The concepts of system, sub-system (or component) and boundaries provide a useful starting point for the engineer to describe and analyse the complexities of real engineering systems. These same concepts also provide a means for breaking down the work of analysis and design into tasks and sub-tasks.

Input and output

Any real system, whether biological, engineered, or corporate, will rarely, if ever, be self-sufficient, i.e. completely cut off from the rest of the world. On investigation, each system is found to interact in some way with other systems, and therefore to be a component in a wider system. A system interacts with other systems, and hence with the rest of the world, by means of **inputs** and **outputs**. These commonly take the form of a flow of matter, energy, or information through the boundaries of the system. The inflow (inputs) represents the influence of the rest of the world on the system; the outflow (outputs) is the influence of the system on the rest of the world. The outputs from one system become inputs to other systems.

A system can be represented schematically by a box with lines indicating input and output, as in Figure 3.1. Inputs and outputs may vary with time,

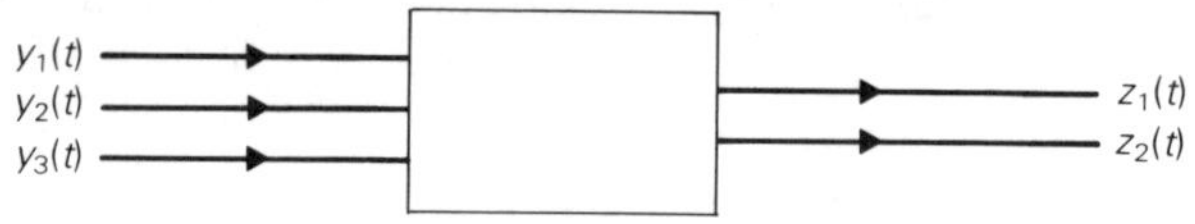

Figure 3.1 A system with input and output.

and are therefore shown in Figure 3.1 as time functions, $y_1(t)$, etc. A large system consisting of components with inputs and outputs can thus be represented schematically with boxes to denote the components and directed lines to indicate the inputs and outputs. The direction of any line defines its relation (input or output) to an attached component. In Figure 3.2, a simple arrangement is shown.

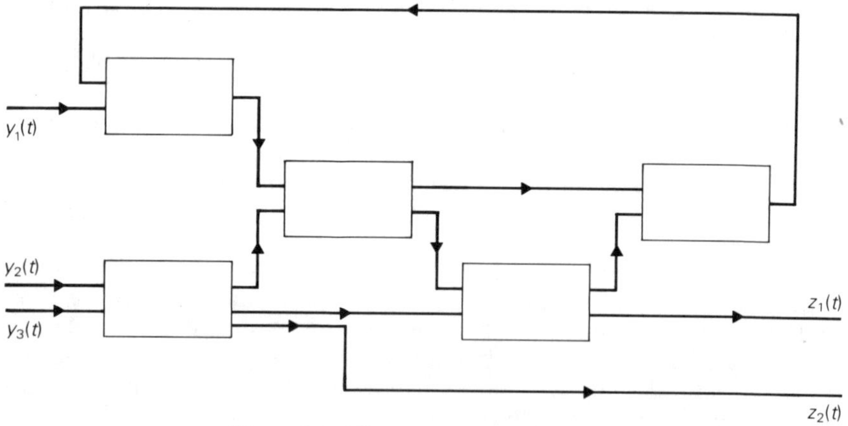

Figure 3.2 The system and components.

Returning to the water supply system as a convenient example, we note that the main inputs and outputs for the components are made up of quantities of water of varying quality which flow into the system, from component to component through the system, and then out again. A schematic representation of the flow of water through a simple system is shown in Figure 3.3. In the case of the water supply system, there will be secondary inputs and outputs, consisting of information on the quality and quantity of water in each component and command signals for water to be pumped from one component to another and to be treated and filtered. These are not shown in Figure 3.3. Other inputs to some of the components will be energy, in the form of electricity or possibly fuel oil, and chemicals for water treatment.

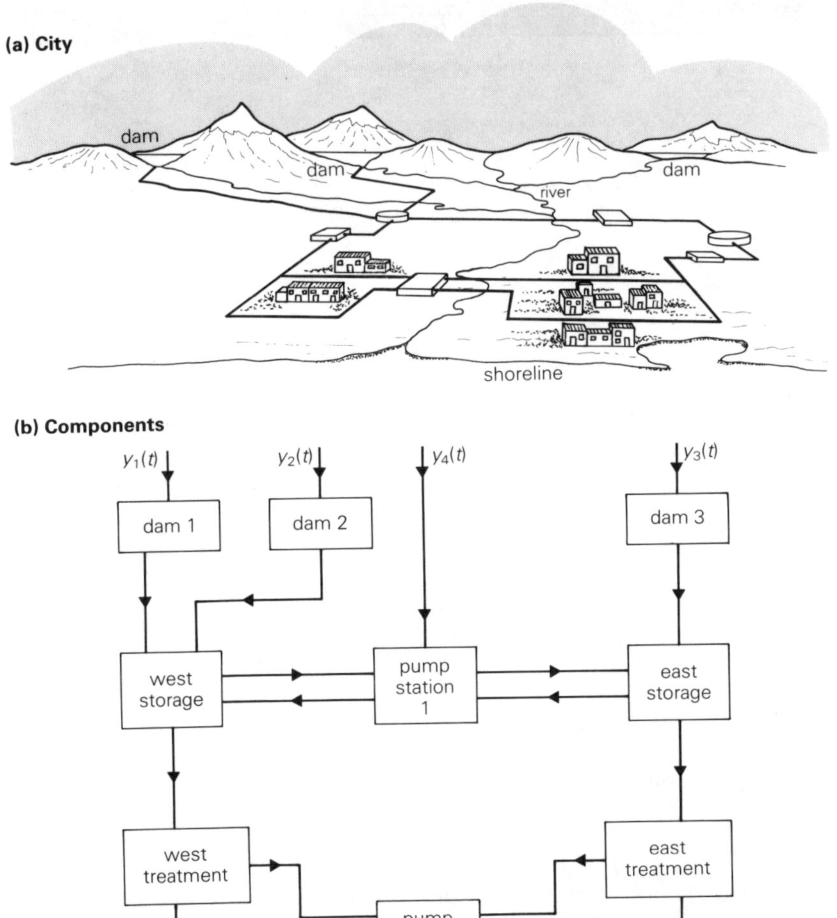

Figure 3.3 A city water supply system.

An example of inputs and outputs which do not fall into the categories of matter, energy, or information occurs in a load-carrying structure, as shown in Figure 3.4. In this case we have chosen the joints as components of the system, the inputs and outputs being the forces which are transferred by the members from one joint to another. An alternative representation could have used the members as components. Similarly, in the finite-element analysis of a massive dam wall, the inputs and outputs across the element boundaries are fields of stresses and strains.

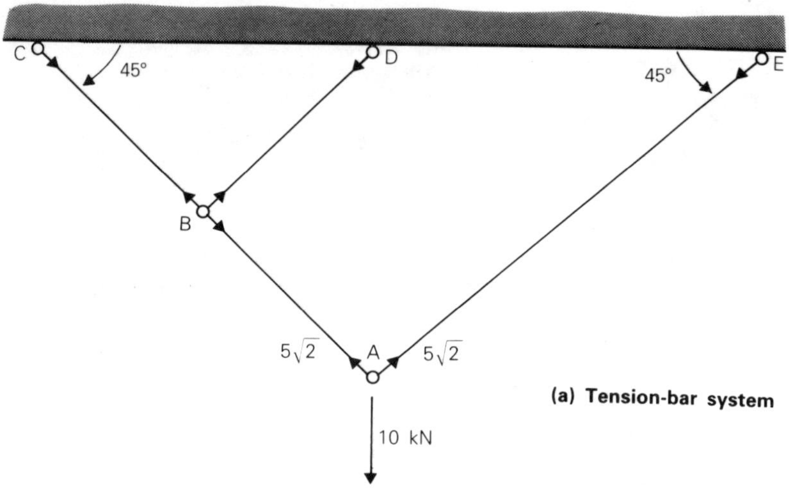

(a) Tension-bar system

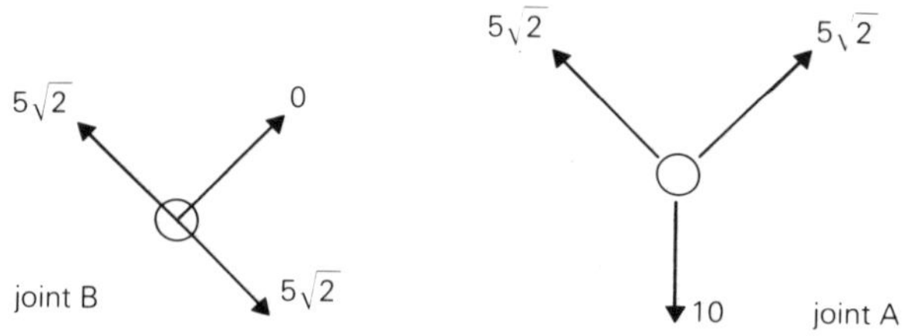

(b) Joints treated as components, with forces as inputs and outputs

Figure 3.4 A tension-bar system.

It has already been emphasized that the designer must pay careful attention to the way in which the components of an engineering system interact, or work together. In other words, the input and output relations of each component with its neighbours must be taken into account in the design of any system. Indeed, the input and output requirements imposed on a component by the other components and by the overall operation of the system determine to a large extent the design criteria for that component.

It has previously been mentioned that a complex system may not display clear-cut natural internal boundaries and components. From a formal point of view, the components separate themselves out from each other if there are clusters of sub-components with relatively few but clear-cut inputs and outputs from one cluster to another. This is shown schematically in Figure 3.5. The 'natural' boundaries thus occur where there are fewest input–output connections among sub-components. When it becomes necessary, for analysis or design, to break a complex system into arbitrary components, the ideal is to choose the boundaries in such a way that the resulting inputs and outputs among the resulting components are as few and as simple as possible.

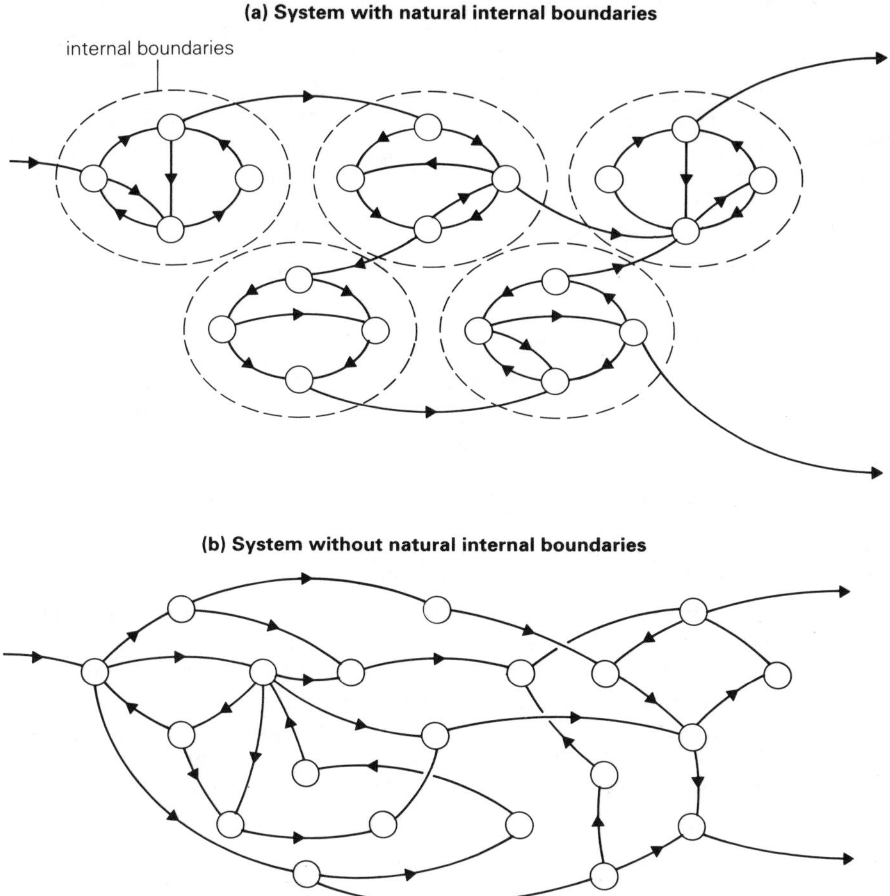

Figure 3.5 Natural boundaries and components.

Black-box components

In order to achieve an overview of the operation of a system it is not necessary to study the detailed functioning within each component. Indeed, by avoiding detail and concentrating on the overall input–output relations among the various components, it is often possible to achieve an improved understanding of the overall operation of a system. When the detailed internal functioning within a component is ignored, and the component is regarded simply as a device which transforms input into output, it may be referred to as a **black box**.

The black-box representation of a component is particularly advantageous when a simplified analysis or design is required, as for example during a feasibility study. An example of a black-box representation would be a simple equation which allows the runoff from a catchment area to be calculated from any given rainfall history. Such a simplified and approximate black-box concept can be used in lieu of a more accurate (but much more complex) study of the rainfall–runoff behaviour by means of a mathematical model or computer simulation.

A highly simplified representation of the complete system can be obtained by lumping closely linked components together, replacing them by black boxes, and representing the important interactions among the black boxes with simple inputs and outputs. At the preliminary stage of analysis or design a detailed, accurate consideration of the functioning of the components is not needed. It may for example be possible to evaluate the input–output relations approximately by experiment or by observation of the system during operation.

Eventually, in the detailed stages of planning and design, it becomes necessary to examine the black-box components in detail and to carry out accurate analysis and design calculations for each individual component.

3.4 Control systems

An engineering system is used to carry out specific tasks and achieve a definite goal. If the goal can be quantified, it may be possible to introduce a device which controls the performance of the system so that it becomes **goal-seeking**. The control device may contain a sensor which in some way detects the present condition, or **state**, of the system. This information can then be sent to another component which compares the actual condition with a desired state, or target condition. A control signal can then be sent to an active device which progressively changes the actual condition, or state, of the system, moving it towards the target state. This arrangement, shown

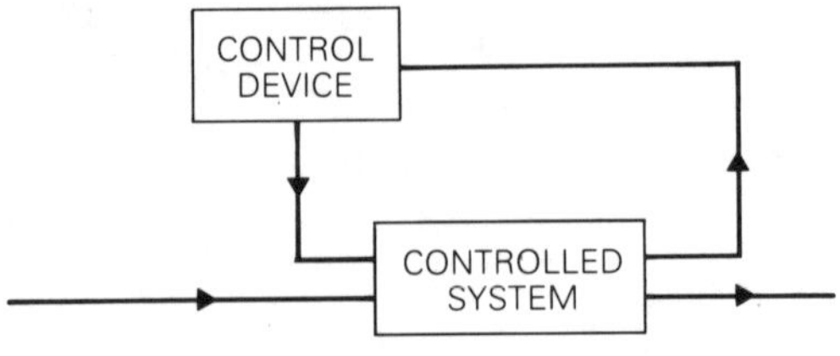

Figure 3.6 The closed-loop control system.

schematically in Figure 3.6, is called a **feedback control system**, or a **closed-loop control system**.

A simple example of a control system is an air conditioner for a room, as shown in Figure 3.7. The desired air temperature in the room is pre-set. This is the target temperature. The thermometer senses the actual temperature in the room, and the control component switches the conditioning unit on when the difference between the measured and target temperatures differ by more than a predetermined margin. Another example of a closed-loop control system is the traffic control lights at a road intersection. Sensors are used to detect and measure the volume of traffic flow into and out of the intersection for all directions. A simple control system calculates the optimal times for flow of traffic in each direction from the measured traffic densities. A control signal is then sent to the lights to control the traffic flows accordingly.

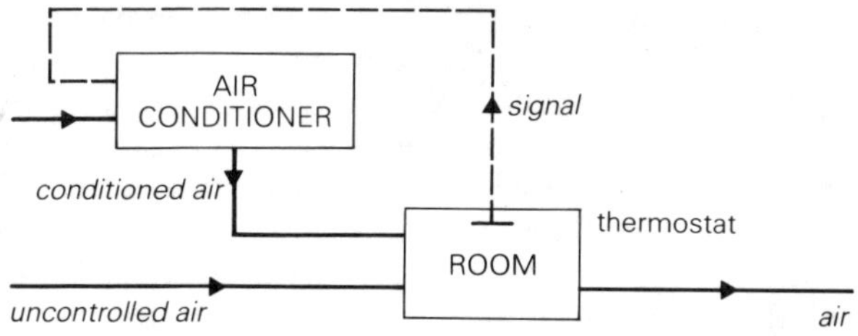

Figure 3.7 An air-conditioned room.

A simpler and less expensive, but also less effective, control system can be achieved if the feedback control signal is omitted. This system, shown schematically in Figure 3.8, is sometimes called, rather illogically, an **open-loop system**. As an example we may consider a set of traffic control lights which operates without the traffic density counters. The timing for the lights to control flow in each direction is in this case pre-set. The pre-set timing may be chosen to allow for normal peak traffic flows at various times of day, but obviously will not handle adequately any unusual occurrences such as weekend sporting events or partial road closure due to accident.

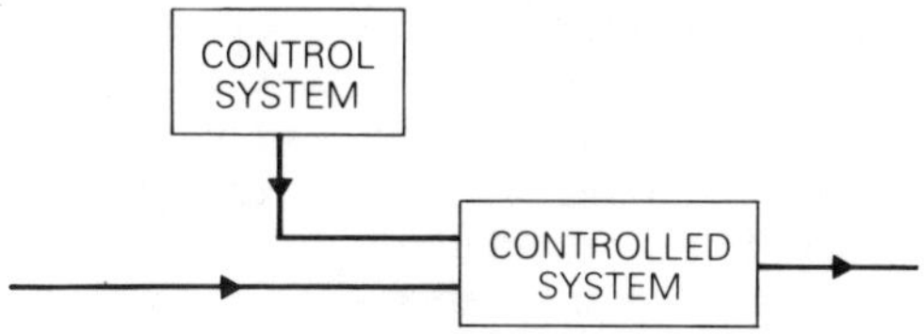

Figure 3.8 The open-loop control system.

3.5 Systems modelling and analysis

System models

A system model is a representation of a physical system which can be used to predict the behaviour of the real system and hence determine how it will function in a given range of operating conditions. The model may be a scaled-down physical version of the real system, as for example when a chemical engineer creates a laboratory model of a proposed chemical plant. A structural engineer may also choose to construct a small-scale physical model of a building in order to observe in a wind-tunnel the occurrence of complex gust loadings. A mechanical engineer may decide to construct a full-scale model of an axle shaft in order to determine fatigue life by repeated load testing. Physical models which have a similar appearance to the original system are sometimes referred to as **iconic models**.

If the physical principles which govern the behaviour of a system are well understood, it will usually be possible to set up mathematical equations to predict the performance of the real system. These equations are referred to as a **mathematical model**. With the advent and widespread use of computers, it has become more and more usual, for reasons of economy, for engineers to use computerized mathematical models to study the behaviour of even quite complex systems. Mathematical modelling is used for the same purpose as physical modelling, namely to study the functioning of the system under prescribed operating conditions, and hence to check or test the adequacy of a proposed design. The role of analysis in the processes of planning and design has already been referred to briefly in Chapter 2.

Even with the most sophisticated and powerful of modern computers, it is still not always possible to develop adequate mathematical models which reproduce with sufficient accuracy the functioning of a complex engineering system. For this reason, physical models still find occasional use. For example, in the study of coastal processes and beach erosion, physical models are used extensively because effective mathematical models have not been developed.

The process of producing a system model, whether physical or mathematical or in the form of a computer program, is referred to as **system modelling**. The subsequent use of a system model to investigate and predict the performance and behaviour of a physical system under a variety of operating conditions is referred to as **system analysis**. We have already seen in Chapter 2 that analysis is an essential step in the iterative design of an engineering system. Analysis is also an important means of studying an existing system to detect inadequacies and to devise improvements.

It is characteristic of all real systems that they undergo changes with time. If the changes occur rapidly, dynamic effects have to be considered in the analysis and design of the system. In this case a **dynamic system model** will be used. On the other hand, if the changes do not take place rapidly, it may be possible to ignore any dynamic effects and consider only the static behaviour, with the use of a **static system model**. In the study of dynamic system behaviour, the inputs and outputs to the system, and also to each component, have to be considered as time-dependent variables, as suggested in Figure 3.1. In the analysis and design of a bridge subjected to highway traffic, consideration has to be given to the dynamic loads which are produced by impact, braking, and even collision; on the other hand, dynamic effects are usually ignored in the design of a small building in a seismically inactive region.

Input–output analysis

The analysis of system behaviour is aimed at predicting the output from a system which occurs as the result of a given input sequence. The input–output relations for a component or system may be thought of as a causal relation, with input regarded as cause and output as effect. The component or system may alternatively be thought of as executing a process, whereby the input is transformed into output. Output can thus be treated as one or more dependent variables, with input as one or more independent variables. If the analysis is of dynamic behaviour, then the variables are all functions of time.

The analysis to determine the input–output relations proceeds typically by progressively breaking the system into simple components and subcomponents. The performance of each component, and in particular the way it transforms its own input into output, can then be studied separately. If the behaviour of a component is too complex for direct analysis, it is broken into components. When the behaviour of the separate components has been studied, the integral effect of the components and hence the overall behaviour of the system have to be determined.

The detailed modelling of the system components and of the integrated system usually involves the application of scientific and engineering principles drawn from the areas of knowledge relevant to the particular physical system.

An important question arises as to the detail with which the modelling of a system should be undertaken. It will be recalled that engineering planning and design are undertaken in a progressive manner, whereby highly simplified designs are initially considered in the feasibility study, and are subsequently refined in later design steps. Ideally, this requires that modelling be undertaken at several levels, of varying accuracy and complexity, for use in the different steps of the design process.

System coefficients

A number of quantities come into the analysis of system behaviour which represent specific properties of the system or of its components. For example, in the truss in Figure 3.4 the lengths of the members and the coordinates of the joints affect the magnitudes of the internal forces. Such quantities are termed **system coefficients**. They should not be confused with the system variables or with the input and output variables. System coefficients are usually fixed in value and represent the known properties of the system. Unlike the state variables and the output variables, they do not change in value in response to changes in the input variables. Material properties and geometric sizes are typical examples of coefficients.

Some system coefficients may be found to change with time, not in response to the input variables, but because the system itself changes. For example, the properties of some engineering materials change slowly with time. The elastic modulus and the compressive strength of concrete both increase slowly with time as the result of complex chemical ageing processes. Both of these properties are important system parameters in the analysis of the behaviour of structures which are made from reinforced concrete and prestressed concrete. Another example of time-varying parameters is when deterioration occurs due to corrosion or abrasive removal of layers of material in a pipe, or scouring and erosion in a stream.

3.6 System state and state space analysis

The concept of system **state** is very useful in analytic studies of the behaviour of dynamic systems. The state of a system is represented by a set of one or more key variables whose values completely define the condition of the system at any instant of time. The key variables are called the **state variables**. Their values at an instant in time comprise the state of the system at that time.

In effect, the state of the system summarizes concisely its entire previous history, and therefore contains sufficient information to allow the future behaviour to be determined for any assumed future set of inputs. The term 'state' has already been introduced informally in the discussion of a control system. Temperature was thus used as the state variable in the case of an air-conditioned room.

When a medical practitioner investigates the state of health of a patient by measuring blood pressure, heartbeat, condition of the eyes, chemistry of the blood and urine (counts of red and white blood cells, levels of cholesterol, triglycerol, uric acid, and sugar), the values of a primitive set of state variables are being determined as a means of establishing the state of a highly complex biological system.

In the analysis of an engineering system, the number of state variables must on the one hand be sufficient to allow the future states of the system to be predicted. On the other hand, if there are too many variables, they will not all be independent. The system state is thus represented by the **minimum** number of independent variables which contain enough information about the system and its past history to allow all future states to be predicted for any given sequence of inputs. In this context, 'past history' refers to values of the state variables at previous instants of time, while 'future states' refers to values of the state variables at all future instants of time.

An analysis of the dynamic behaviour of a system can be carried out by deriving a set of equations which allow the incremental or differential changes in the values of the state variables to be evaluated at succeeding time instants. Such equations are known as the **equations of state**. The range of possible values of the state variables can be thought of as defining a mathematical space. The state of the system at any instant is represented by a point in this space, which is known as **state space**. This type of analysis and representation is often referred to as **state space analysis**.

It is beyond the scope of this book to go into the details of state space analysis of dynamic systems. Nevertheless, various very important system concepts, including linearity, non-linearity, stability, and instability, follow on clearly and unambiguously from the state-space approach to systems analysis. Various texts are available for further information. Although dated, the text by Elgerd (1967) gives a very simple and clear introduction in the context of electrical engineering, while a classic, entertaining, non-mathematical treatment of state concepts is provided by Ashby (1956). A simple and brief example of the state space approach to dynamic analysis is provided for readers with some mathematical background in the Appendix to this chapter.

3.7 System simulation

A computer program is frequently required to carry out the detailed calculations which follow on from an accurate analysis of the behaviour of a complex dynamic system. A **simulation** of the system is said to be made when values of the state variables or output variables are determined over a specified period of time for a prescribed set of input conditions and system coefficients. A simulation of system behaviour would normally be carried out in the detailed stage of design and planning for a large engineering project. System simulation thus serves as a form of detailed systems analysis.

System simulations are sometimes repeated a large number of times with different values of the system coefficients and inputs, in order to observe the effect on system behaviour of small variations in the coefficients and inputs. The term **sensitivity analysis** is sometimes used for such studies.

3.8 Optimization

An engineering system has to have a clearly defined goal in order for it to be designed and realized. A clearly defined objective is likewise a prerequisite for the planning of an engineering process. In either situation, the task of the designer is to achieve the best possible solution, given the constraints which must be satisfied.

By the time the detailed planning and design work is under way, a specific approach has been decided on and the main task is to work out the design and planning details. In the detailed stages, optimization techniques can sometimes be used to find, for example, the values of the system parameters which maximize the performance of the system. The term **optimization** refers to techniques (usually mathematical), whereby the details of the system are adjusted in such a way as to achieve the best possible system performance, or alternatively the cheapest design which achieves a specified performance level.

In optimizing a system or process it is imperative that the measures of effectiveness be clearly and unambiguously defined. Otherwise the concept of optimization is meaningless. In mathematical procedures, a measure of effectiveness is expressed as a mathematical function of the inputs and outputs of a system, or possibly of the system state. It is then called the **objective function**. The system parameters or coefficients are the quantities which are adjusted to produce an optimal system. The design or planning constraints are also expressed in mathematical terms, often as inequalities.

Various mathematical methods are available for use in the optimization of a system or process. These include linear, non-linear, and dynamic programming, and the parametric investigation of the design parameters. Informal techniques include optimization by search, optimization by experiment and system simulation. Optimization techniques are discussed in some detail in Chapters 7 and 8.

3.9 Error of sub-optimization

It is interesting and important to note that optimization of the overall system will usually *not* be achieved by independent optimization of each individual component. Indeed such an approach can lead to serious diseconomies and malfunction of the overall system. The optimization of an individual component can be detrimental to the overall system.

The term **error of sub-optimization** is used to describe the optimization of a component to the detriment of overall system performance.

Sub-optimization errors occur for example in the structural design of buildings when too much emphasis is placed on a minimum weight solution, on the (usually mistaken) assumption that this will lead to minimum cost. The problem is that structures designed for minimum use of materials can be very costly to construct. The additional construction costs can far outweigh the small savings in materials in a building. On the other hand, minimum weight is a much more important objective in aircraft design.

In the design of an aircraft, overall performance is the objective. If the engines are designed to achieve maximum performance without due regard to their fitting into the airframe and working in conjunction with other components, the result will be sub-optimal so far as the aircraft is concerned. The design of a complex system is always a compromise, whereby each component is 'tuned', not in itself to perform optimally, but to contribute to the optimal performance of the overall system.

Sub-optimization errors are prone to occur in large engineering projects when insufficient contact is maintained among different groups working independently on separate aspects which must eventually be brought together. One of the prime tasks of project planning and management is to minimize the occurrence of sub-optimization errors.

Appendix: equations of state

To illustrate the development of a set of equations of state, we consider the dynamic analysis of the two-storey frame shown in Figure 3.9. All system variables and coefficients are defined in Table 3.1. The building is subjected to horizontal loads due to wind or possibly earthquake.

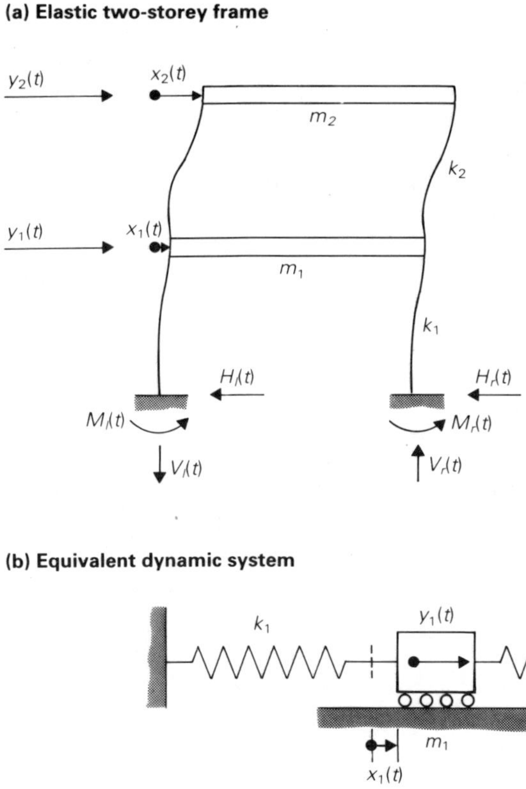

Figure 3.9 State-space analysis of a two-storey frame.

In the preliminary design calculations the floors are considered to be stiff in comparison with the columns and hence to move laterally under the influence of the horizontal dynamic forces, but not to bend. The forces $y_1(t)$ and $y_2(t)$ act at the floor levels and vary with time. Horizontal movement of the floors occurs because of the flexibility of the supporting columns.

In order to build a state model of this system we must choose appropriate state variables to represent the condition of the system. The horizontal displacements at the floor levels, $x_1(t)$ and $x_2(t)$, are clearly appropriate. However, these two variables alone are not sufficient to specify the state of the system: initial values of the displacements at some starting time t_0, i.e. $x_1(t_0)$ and $x_2(t_0)$, do not contain enough information to allow all future states

Table 3.1 System variables and coefficients for two-storey frame.

Input variables
$y_1(t)$ (N): force applied at first-floor level
$y_2(t)$ (N): force applied at second-floor level

State variables
$x_1(t)$ (mm): displacement at first-floor level
$x_2(t)$ (mm): displacement at second-floor level
$x_3(t)$ (mm/s): velocity at first-floor level
$x_4(t)$ (mm/s): velocity at second-floor level

Output variables
$z_1(t)$ (mm): displacement at first-floor level ($= x_1(t)$)
$z_2(t)$ (mm): displacement at second-floor level ($= x_2(t)$)

System coefficients
m_1 (kg): mass of first floor
m_2 (kg): mass of second floor
k_1 (N/mm): stiffness of ground-floor columns in resisting first-floor displacement of $x_1(t)$
k_2 (N/mm): stiffness of first-floor columns in resisting second-floor displacement of $(x_2(t) - x_1(t))$

to be evaluated for any given input history $y_1(t)$ and $y_2(t)$. In fact, we need to include two further variables. These might be chosen as the velocities of the floors,

$$x_3 = \frac{dx_1}{dt} = \dot{x}_1 \tag{3.1}$$

$$x_4 = \frac{dx_2}{dt} = \dot{x}_2 \tag{3.2}$$

If initial values of these four state variables are known at the initial time instant t_0, it is then possible to determine, from a dynamic analysis of the simplified system shown in Figure 3.9(b), the subsequent states of the system for any specified history of the applied forces.

The equations of state can be derived from simple dynamic principles. In Figure 3.10 the top floor is represented as a simple point mass which is accelerating under the influence of the resultant of the two forces acting on it. One of these forces is the external load $y_2(t)$; the other is exerted by the columns in the second storey. The stiffness of these columns is k_2 in N/mm. (That is to say, a horizontal force of k_2 N when applied at the top-floor level produces a lateral movement of the second floor, relative to the first floor, of 1 mm.) By setting force equal to mass times acceleration we obtain

$$y_2(t) - k_2\,[x_2(t) - x_1(t)] = m_2\,\ddot{x}_2(t) \tag{3.3}$$

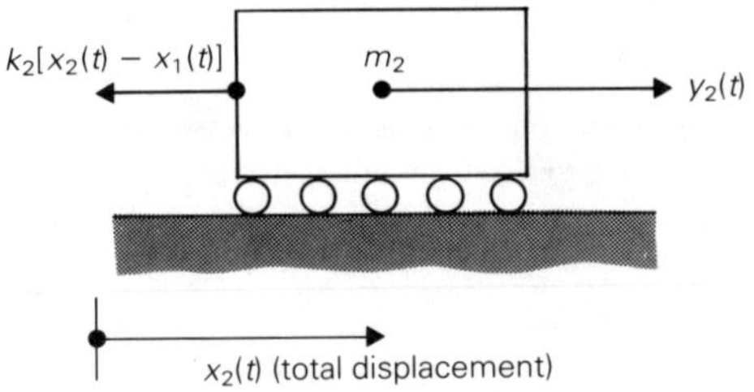

force = mass × acceleration

$$y_2(t) - k_2[x_2(t) - x_1(t)] = m_2\ddot{x}_2(t)$$

$$\dot{x}_4(t) = \ddot{x}_2(t)$$

Figure 3.10 Dynamics of the top floor.

Noting also the definition of the fourth state variable,

$$x_4(t) = \dot{x}_2(t) \tag{3.4}$$

we obtain the following state equation:

$$\dot{x}_4(t) = \frac{1}{m_2}\,[y_2(t) - k_2\,[x_2(t) - x_1(t)]] \tag{3.5}$$

Similar considerations apply to the mass at the first-floor level. A set of four linear first-order differential equations is thus obtained in which the derivative of each state variable is expressed as a function of the inputs, coefficients and current values of the state variables. These are the equations of state for the system, expressed in normal form. We have:

$$\dot{x}_1(t) = x_3(t) \tag{3.6}$$

$$\dot{x}_2(t) = x_4(t) \tag{3.7}$$

$$\dot{x}_3(t) = \frac{1}{m_1}\,[y_1(t) - k_1x_1(t) + k_2[x_2(t) - x_1(t)]] \tag{3.8}$$

$$\dot{x}_4(t) = \frac{1}{m_2}\,[y_2(t) - k_2\,[x_2(t) - x_1(t)]] \tag{3.9}$$

Various numerical and analytic techniques can be used to solve the equations and hence obtain the values of the state variables at a sequence of time instants, given an initial starting state and predefined inputs. However, we are not concerned here with the mathematical methods, and we shall not pursue the example further.

References

Ashby, W. R. 1965. *An introduction to cybernetics*. London: Methuen.
Ashby, W. R. 1965. *Design for a brain*. London: Methuen.
Elgerd, O. I. 1967. *Control systems theory*. Tokyo: McGraw-Hill Kogakusha.
McLoughlin, J. B. 1969. *Urban and regional planning. A systems approach*. London: Faber & Faber.

Problems

3.1 Consider the problem of the planning and design of a new domestic airport for a small coastal city.

(a) Identify and describe briefly the main components (sub-systems) of the airport. Also identify and describe briefly the sub-sub-systems in several of the main components.
(b) What might be considered to be the main inputs and outputs of the airport?
(c) In planning the airport to function effectively, which airport processes need to be considered carefully and subjected to quantitative analysis?
(d) List some of the hardware components which need to be designed, and indicate what the design criteria might be for each component.
(e) Which variables would you use to specify the state of the airport at any particular time?

3.2 Repeat the steps in Problem 3.1 for

(a) a ten-storey office building
(b) a water supply system and sewage system for a small township.

3.3 Identify a control system in (a) an automobile and (b) a house. Describe briefly the components of the system and their function.

4 Creative aspects of planning and design

This chapter deals with the role of creativity in planning and design. Creativity is required for the generation of concepts during the feasibility phase of a project. The quality of the final design or plan critically depends on the ability of the engineer to use experience and imagination to identify possible solutions to the problem.

4.1 Creativity

Before considering techniques which may help in the generation of alternative concepts it is useful to examine the nature of creativity and the factors which can enhance or discourage creative thought.

Creativity may be defined as 'the origination of a concept in response to a human need – a solution that is both satisfying and innovative' (Beakley and Leach 1972).

A clear distinction should be made between analytical thought and creative thought.

Analytical thought requires deductive reasoning. It is applicable to problems which have a few 'correct' answers or possibly only one. Analytical thought is a convergent process which starts with many ideas and leads to the identification of a single solution to the problem. **Creative thinking**, on the other hand, tends to be divergent and produces many ideas for solving a problem. The complete design process involves both creative and analytical thinking: first, a number of possible solutions are generated, then these are progressively reduced as they are analysed and evaluated in more detail until a single, satisfactory solution is obtained.

It is an unfortunate feature of many engineering degree courses that analytical thinking is overwhelmingly emphasized at the expense of creative thinking, and insufficient encouragement and practice is given to students in the development of new, novel, and innovative ideas.

Every individual has the ability to be creative. It usually takes considerable self-discipline to engage in effective creative thought and not to allow oneself to drift into 'daydreaming' or 'stargazing'. Many people in our society are not nearly as creative as they could be, because either their educational development or their work environment strongly discourages creativity.

In applying scientific knowledge to meet the needs of society the engineer has a unique opportunity to be creative in a manner matched by few other professionals. As Theodore Von Karman said, 'scientists study the world as it is, engineers create the world that never has been' (Beakley and Leach 1972).

In fact, it is not only desirable but necessary for engineers to be creative in their work. The success of an engineering project often depends on the engineer's ability to develop new, innovative, and effective solutions.

Of course not all engineering activity is creative. In fact a large proportion of an engineer's time may be spent in non-creative activities. However, most engineering jobs provide some opportunity for creative thinking. It should be borne in mind that the development of an effective scheme to reorganize the office administration may be just as creative as the design of an improved model of racing car.

4.2 Factors that influence creative thinking

Psychologists recognize that individuals differ considerably in their ability to engage in creative thinking. In fact there are a large number of factors which may determine the creative potential of an individual at any time. It has been suggested that the following factors are important (Beakley and Leach 1972, Hicks and Gullett 1976):

- personal attributes
- motivation
- emotional stress
- education and experience
- social and cultural barriers

We now discuss each of these in turn.

PERSONAL ATTRIBUTES

There appears to be some correlation between intelligence and creativity, although many very creative people are of average intelligence according to traditional IQ tests. It has also been suggested that good health assists in creative thinking. However, there are many examples of creative people who had physical and mental afflictions, for example Beethoven and Van Gogh.

To engage in creative thought an individual must be able to accept conflicting ideas and often disorder. She must be able to accept that there are often many possible solutions to a problem, some of these being totally unconventional. An inquiring mind and constructive self-criticism are often the hallmark of a creative thinker.

MOTIVATION

Creative thinking usually involves considerable application and effort by the individual concerned. Ideas do not just come 'out of the blue', but may require a large amount of background reading and digestion. Strong motivation in terms of interest in one's work and a desire to succeed have a positive bearing on creative thinking. Conversely, an engineer who feels that his ideas are unlikely to be adopted by his superiors will not be motivated to engage in creative thought.

EMOTIONAL STRESS

Every individual experiences various emotions at one time or another. These include anger, joy, fear, guilt, and anxiety. The ability to engage in creative thought depends on the level of emotional stress felt at the particular time.

Many people are more creative when subjected to low levels of stress. Such people have their most creative thoughts when engaged in activities such as bathing, listening to music, or gardening. For example, the Greek mathematician and engineer Archimedes is said to have discovered his famous principle of hydrostatics while bathing. On the other hand, excessive stress levels usually act to inhibit or totally suppress creative thoughts. Such high stress levels may be caused by the death of a loved one, divorce, separation, changing jobs, becoming a parent, or financial difficulties.

In some cases the working environment may place an individual under stress and thereby reduce his or her creativity. For example, an organization where there is a large degree of conflict between individuals is not conducive to creative thought.

Nevertheless, some individuals work better under stress and produce their most creative work when under the pressure of deadlines or when experiencing intense emotional stress.

EDUCATION AND EXPERIENCE

It is generally believed that the more knowledge that is possessed by an individual, the more likely it is that the individual will produce creative ideas. Education adds to our knowledge base and so can enhance the potential for creativity. In a similar way, experience also adds to our knowledge base and so can enhance creative thinking. In a sense education is a compilation of the experience of others and so the two may be seen to be complementary.

However, education and experience can also act to repress creative thinking. If education does nothing more than pass on a set of standard solutions to problems and does not encourage the individual to think creatively and identify new solutions, it may act as a negative influence upon creative thinking. Similarly, experience which leads to the continuation of traditional practices without adaptation to new materials and technology works against the development of creativity.

SOCIAL AND CULTURAL BARRIERS

Our society has erected many barriers to creative thinking. Most individuals find it difficult to cope with change and therefore tend to oppose new ideas. There is also a tendency in scientific circles to gain kudos by ridiculing the ideas of others. Such activities inhibit the expression of new or unusual ideas and so limit creative thinking. In fact, the fear of ridicule is probably the biggest single barrier to creativity.

Each individual has inbuilt social and cultural barriers which suppress ideas considered to be too bizarre or stressful for that individual to cope with. For example, the survivors of a plane crash in the Andes Mountains in 1972 had no food (Read 1974). They knew it would take weeks before they could be rescued. Some of them thought of eating the bodies of those who had died in the crash. Despite the strong cultural barriers to this practice a number of people (including devout Catholics) managed to carry it out and survived as a consequence. Others, who could not bring themselves to cannibalism, grew weak and died in the extreme cold.

4.3 Techniques for stimulating ideas

As creativity is an essential component in all planning and design it is important for engineers to make maximum use of their personal creative abilities.

Various techniques have been developed to assist in the generation of ideas. The techniques may be grouped into the following two categories:

- **analytical techniques**, such as attribute listing and morphological synthesis
- **free association techniques**, such as brainstorming and the Nominal Group Technique.

Analytical techniques utilize a logical analysis of all the components of the problem. These techniques are usually best applied by an individual working alone.

Free association techniques are group activities designed to encourage the free flow of ideas. They have been developed to tap the creative potential of the subconscious part of the human brain. In the absence of these techniques, creative (and often irrational) ideas from the subconscious tend to be suppressed because they are thought to be stupid or because they cross social and cultural barriers. Free association techniques encourage all ideas to be expressed, regardless of how fantastic they may seem, by not allowing any criticism of the ideas within the group. In fact, by promoting the interaction of a group of people there may be positive encouragement, or even competition to produce new ideas.

Attribute listing

The technique of 'attribute listing' is most applicable in cases where there is an existing solution and ways to improve on it are being sought. The steps involved are:

- list all of the components or elements of the existing solution
- list all the attributes of the product and its components (Attributes include weight, size, colour, shape, material, and so on.)
- systematically change each attribute of the product and its elements in every conceivable way

As an example, let us consider developing improved designs for a bicycle using attribute listing.

The components of the existing product are:

wheels	chain
tyres	brakes
frame	gears
handlebars	seat
pedals	mudguards

The attributes of each of these components are listed and systematically changed. Take wheels, for example: possible variations are given in Table 4.1. A similar procedure is followed for each component.

Table 4.1 Changes in the attributes of wheels.

Attribute	Variations
Two in number	Three, four, one, none (e.g. caterpillar tread)
Steel rims	Fibreglass, plastic, or aluminium rims
Steel spokes	Fewer spokes, plastic spokes, solid disc
Chrome plated	Painted, anodized, rusted
65 cm in diameter	Larger, smaller, two different sizes
Roller bearings	Nylon bearings

Now, combinations of the variations in each attribute can be developed to produce a series of design concepts.

Morphological synthesis

This technique is also called **matrix analysis** (Allen 1962, Beakley and Leach 1972). It is based on the generation of unusual combinations of system functions. The steps involved are:

- describe the problem
- list the major system parameters
- list the alternative ways of satisfying each system parameter
- draw up a matrix with each system parameter as one dimension
- consider all possible combinations of ways of satisfying each system parameter.

For example, it is desired to establish an innovative mode of public transport in a major city. Morphological synthesis will be used to generate alternatives.

Going through the steps listed above, in order:

- the problem is to find an innovative mode of public transport

- the major system parameters might be: carriageway, energy source, method of support
- alternatives for each system parameter are:

  Carriageway – rail
  – road
  – monorail
  – water
  Energy source – liquid fuel
  – electricity
  – gas
  – solar
  Method of support – wheels
  – air
  – magnetic
- The combinations of parameters are depicted in Figure 4.1.

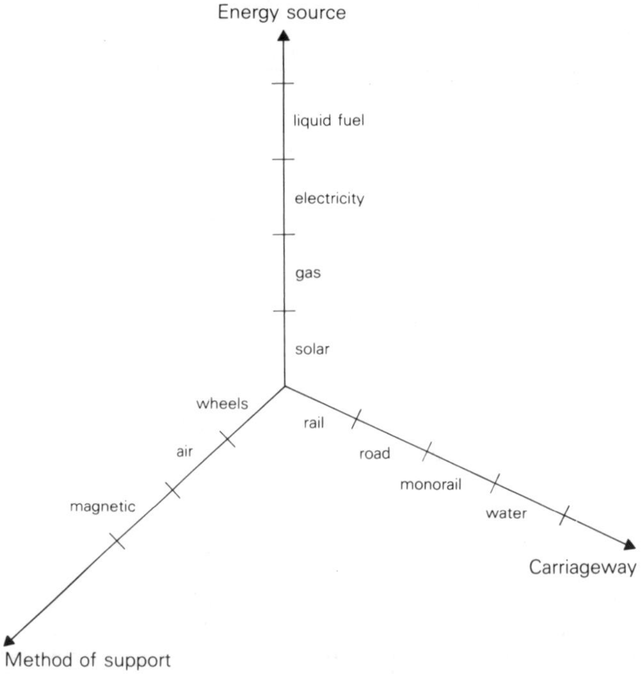

Figure 4.1 Combinations of parameters for a public transport system.

A total of 48 combinations are depicted. These include:

- an electric-powered monorail with magnetic support (or suspension)
- gas-powered vehicles travelling over water with air suspension (i.e. hovercraft)
- solar-powered buses with wheels travelling on the road

Brainstorming

The brainstorming technique has been widely applied in industry and government (Osborn 1963, Rawlinson 1981). A brainstorming session usually involves 6–12 people. It is desirable to include people with a broad knowledge as well as experts in the field of the particular problem under study. Two officials are required, a chairperson and a recorder. The chairperson introduces the problem and must strictly enforce the rules. It is her responsibility to maintain a constant flow of ideas.

The recorder records *all* ideas presented. They are best recorded on a blackboard or large sheet of paper which is visible to all participants. This ensures that visual stimulae reinforce the spoken word.

The rules of a brainstorming session are:

- all ideas must be recorded
- no idea may be criticized or evaluated during the session
- it is permissible to build on, expand, or combine the ideas of others
- the emphasis is on the *quantity* of ideas produced. It has been shown that the *quality* of the final solution depends on the number of ideas presented during the session

At the start of the session the chairperson should introduce the problem and answer any questions which are needed to clarify it. It is often better for members of the group to be informed of the problem several days before the session. The session would then normally last about 30–45 minutes after the introduction.

The ideas generated in a brainstorming session should be considered for detailed evaluation several days later. This gives time for the participants to recover and for the ideas to gel. In contrast to the brainstorming session, evaluation is better when carried out without consultation between participants. One mechanism is to give all participants a list of the ideas generated and ask them to select, without consultation, 10% of them which they believe are worth more detailed examination. The results are collected and the better ideas identified. A smaller team would then start to 'flesh out' the options and test their feasibility.

Often, in the course of a brainstorming session the flow of ideas can wane and there is a prolonged silence. A 'trigger' is a very useful device to restart the flow of ideas. A number of types of triggers have been suggested. These include:

- a checklist
- wildest fantasy
- juxtaposition

One checklist is provided by the acronym SCAMPER and is listed in Table 4.2.

The use of 'wildest fantasy' as a trigger encourages all participants to think of the wildest idea. Some of these may lead to other more practical solutions.

Table 4.2 The checklist SCAMPER.

Substitute	Who? What? Other material?
Combine	Units? Purposes? Ideas? Other place?
Adapt	What else is like this? What can I copy?
Modify	Change meaning? Shape? Colour?
Put to other uses	
Eliminate	Parts of
Rearrange	Interchange components; other layouts

'Juxtaposition' involves the following steps:

- select three words at random
- discover a connection between each word and the problem being studied

(The best words to use are familiar ones, such as water, hair, mouth, meat, cup, pen, etc.)

For example, in trying to suggest ideas for the design of a land-based vehicle which does not use fossil fuels a group has run out of ideas. They decide to use juxtaposition as a trigger, and select the words pencil, water, and chicken. The word pencil leads to the suggestions of using wood as a fuel, making the vehicle long and thin to reduce drag, using graphite as a fuel or lubricant, and a vehicle which walks on stilts.

The nominal group technique

The nominal group technique is somewhat different from brainstorming as it encourages all participants to contribute in a quiet, non-competitive way (Deason and White 1984). The group is usually composed of between five and nine members. The steps involved are as follows:

- individuals generate ideas in writing
- all ideas are recorded in a round-robin fashion
- ideas should be clarified in discussion
- a series of voting rounds are held to select a few recommended solutions

In the first step the leader explains the problem and answers any points of clarification on it. All participants are then given a written statement of the problem. Each participant is then asked to independently write down ideas for solving it. There should be no discussion during this generation process.

After a reasonable amount of time for recording ideas, each participant in turn is requested to contribute one idea to the group. These are recorded in writing on a blackboard or flip chart. The opportunity to each contribute one idea passes around the group until all ideas have been recorded.

There is then an opportunity to clarify the ideas. Each one is discussed in order to ensure that all members of the group are clear on its meaning. The leader must ensure that the discussion does not get out of hand. Similar ideas may also be combined at this stage.

Participants are then asked to independently select a set of about five or six ideas from the list and record these as being the best ones. A short list for the group is then obtained by collecting the 'best' ideas recorded by all individuals.

If a smaller list is desired this can be obtained by further voting. It is usually better to separate each voting round by a brief discussion in order to clarify why people favour particular ideas.

The advantages of the Nominal Group Technique are that it draws on each participant's creativity without allowing one or two individuals to dominate. Discussion is kept to a minimum, and evaluation involves the combined judgement of all individuals.

References

Beakley, G. C. & H. W. Leach 1972. *Engineering, an introduction to a creative profession*, 2nd edn. New York: Macmillan.

Deason, J. P. & K. P. White Jr. 1984. Specification of objectives by group processes in multiobjective water resources planning. *Water Resources Research* **20**(2), 189–196.

Dickerson, S. L. & J. E. Robertshaw 1975. *Planning and design*. Lexington: D. C. Heath.

Goicoechea, A., D. R. Hansen & L. Duckstein 1982. *Multiobjective decision analysis with engineering and business applications*. New York: John Wiley & Sons.

Hicks, H. G. & C. R. Gullett 1976. *The management of organizations*, 3rd edn. Tokyo: McGraw-Hill Kogakusha.

Kagan, J. & E. Haveman 1980. *Psychology. An introduction*, 4th edn. New York: Harcourt Brace Jovanovich.

Kemper, J. D. 1985. *Introduction to the engineering profession*. New York: Holt, Rinehart and Winston.

Osborn, A. F. 1963. *Applied imagination*. New York: Scribners.

Rawlinson, J. G. 1981. *Creative thinking and brainstorming*. Aldershot: Gower.

Read, P. P. 1974. *Alive: The story of the Andes survivors*. Philadelphia: J. B. Lippincott Co.

Woods, D. R. 1985. *The McMaster Problem Solving Program*. Unit 8. Creativity. Unpublished notes. Department of Chemical Engineering, McMaster University, Hamilton, Ontario.

Problems

4.1 Use the attribute listing technique to come up with alternative design concepts for the following:

(a) an electric shaver
(b) a chair
(c) a lawnmower
(d) a kite

4.2 Set up and conduct a brainstorming session to generate ideas for the following problem:

You are a member of a team of six engineers who wish to enter a competition to design a vehicle capable of transporting two adults from Perth to Sydney in the minimum amount of time. The direct distance is approximately 3200 km (see Figure 4.2). The propulsion of the vehicle must not involve the combustion of liquid, solid, or gaseous fuels. The vehicle must be designed and constructed within 12 months or less and may not cost more than $100 000 to build.

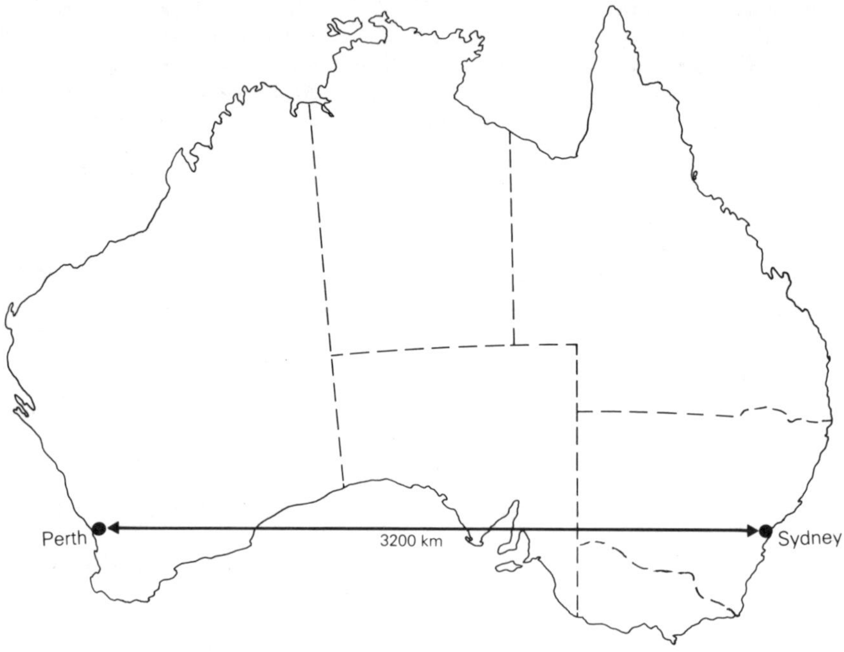

Figure 4.2 A map of Australia, showing Perth and Sydney.

4.3 Design and construct a bridge out of a packet of spaghetti (500 g) and a reel of cotton. The bridge is to be supported on piers with a span of 300 mm. The bridge must be able to carry a load of 500 g placed at any point between the supports. The least cost solution is the one which uses the least weight of materials.

4.4 Devise a protective mechanism for a raw hen's egg, which is to be dropped a distance of 3 m on to a concrete pavement from a window.

The materials available for the exercise are 20 sheets of A4 writing paper and a small container of paper glue. The least cost solution is the one with the least weight of paper and glue.

5 Economic evaluation

This chapter deals with the economic evaluation of engineering projects. The purpose of economic evaluation is to identify the benefits and costs of a project and hence to determine whether it is justified on economic grounds. It is usually necessary to discount the future benefits and costs of a project in order to make comparisons in terms of present values. Various economic criteria have been proposed for the comparison of projects. Each of these has its merits, although the best on theoretical grounds is **net present value**. *The basis for estimating benefits and costs also needs careful consideration.*

5.1 Introduction

An engineering project involves the transformation of limited resources into valuable final products or outputs. For example, the construction of a dam involves the use of resources such as concrete, steel, human effort, and machine time. The dam is used to produce outputs such as water for domestic and industrial purposes, flood control, and recreation.

The engineer must be concerned not only with the technical feasibility of the project. (i.e. will it work?) but also the economic feasibility. Economic evaluation is aimed at assessing whether the value of the final products exceeds the value of the resources used by the project. The values of the outputs when measured in economic terms are called the **benefits** of the project. Similarly the values of the resources used in its construction and maintenance when measured in economic terms are called the **costs** of the project.

In the private sector, benefits and costs are measured by cash flows into and out of the firm, respectively. That is, a conceptual boundary is drawn around the firm and benefits, and costs are represented by cash flows across this boundary.

For public sector projects, benefits and costs must be considered for society as a whole. In this case, benefits and costs are not necessarily associated with cash flows. For example, the benefits of a public transport system are not necessarily measured by the revenue which it generates. The government may choose to run the system at a loss in order to provide cheap transport to low-income groups. In this case the benefits to the users of the system would probably exceed the revenue generated by it. If, for example, public transport were offered free of charge, few people would argue that the benefits to society are zero.

Sometimes there is a different perception of benefits and costs at various levels of government. For example, the federal government may provide a subsidy of 50% of the capital costs of new sewage treatment plants. When a state government is evaluating a particular plant it would consider this subsidy as a reduction in the cost of the project. On the other hand, the federal government would consider the subsidy as merely a transfer payment from one branch of government to another and therefore not a true benefit or cost to society as a whole. In fact, the federal government may require the state to provide economic justification for the project in its (i.e. the federal government's) terms without including the subsidy effect.

5.2 The time value of money

The costs and benefits (or revenue) of an engineering project usually occur over a long time period. For example, consider the construction of a new freeway. A typical time stream of benefits and costs is illustrated in Figure 5.1. Costs will be high during the construction phase which may last for three or four years. Annual maintenance and repair costs will be low initially, but will increase due to ageing of the pavement, bridges, and other components. The benefits to road users will be primarily due to savings in vehicle operating costs, savings in travel time and a reduction in the number of accidents in relation to the pre-existing road network. These benefits would normally increase with time du› to increasing volumes of traffic using the freeway.

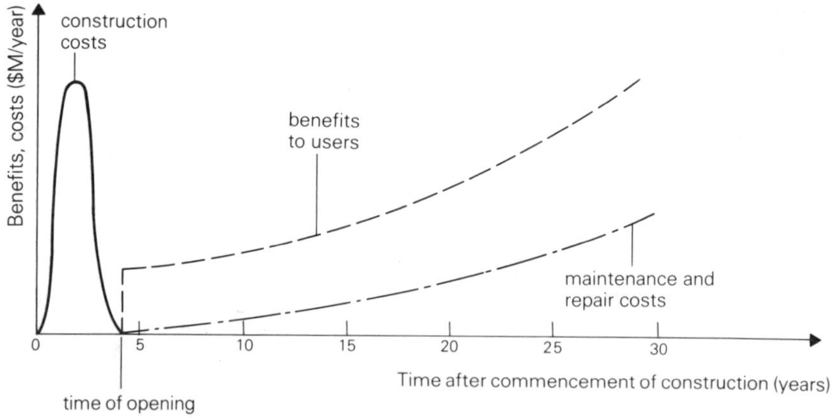

Figure 5.1 The time stream of benefits and costs for a freeway.

In carrying out an economic evaluation of such a project it must be recognized that benefits or costs incurred in ten years' time cannot be directly compared with those incurred in the current year. Given the choice between $1000 now and $1000 in ten years' time, very few people would choose the latter. In the first place, the effects of inflation are such that fewer goods and services could be purchased with the future sum than at present. However, even in the absence of inflation, human nature is such that there is a preference to consume goods now rather than later, and to postpone costs if possible. This is clearly evidenced by the fact that many individuals are willing to borrow money at an interest rate which exceeds the rate of inflation. Therefore in carrying out economic evaluation it is necessary to discount future benefits and costs in order to make them directly comparable with benefits and costs incurred now. This is carried out using discounting formulae derived from considerations of compound interest.

5.3 Discounting formulae

In the development of discounting formulae the following assumptions are made:

- there is a single interest rate which applies to both lender and borrower
- interest is applied in a compound fashion, i.e. the interest paid in any time period is based on the principal plus the interest from any previous time period(s).

Present worth of a lump sum

Consider a sum $\$P$ deposited in a bank account today at an interest rate of $i\%$ per annum. The total sum (principal plus interest) held in the account after n years is denoted by $\$F_n$. After one year the total balance including interest is

$$F_1 = P(1+i) \tag{5.1}$$

After two years, interest is paid on the total sum F_1, so that F_2 is given by

$$F_2 = P(1+i)(1+i) = P(1+i)^2 \tag{5.2}$$

Clearly after n years the total sum is given by

$$F_n = P(1+i)^n \tag{5.3}$$

Equation (5.3) gives the future sum, $\$F_n$, that is equivalent to a present amount, $\$P$. By convention, the factor $(1+i)^n$ is called the **compound amount factor** and is denoted by $(F/P,i\%,n)$. Values of this and the other compound interest factors are widely tabulated or can be readily determined using a pocket calculator. By transposing Equation (5.3) the following expression is obtained:

$$P = \frac{F_n}{(1+i)^n} \tag{5.4}$$

In this case, $\$P$ is the amount of money which must be invested now to give a compound sum of $\$F_n$ in n years' time. Another interpretation of Equation (5.4) is that $\$P$ is the present worth of $\$F_n$ received in n years time. The factor $1/(1+i)^n$ is called the **present worth factor** and is denoted by $(P/F,i\%,n)$. It is the reciprocal of the compound amount factor.

Example 5.1 What is the present worth of $1000 received in 10 years' time if the interest rate is 10% per annum compound?
Solution Using Equation (5.4),

$$P = 1000/(1+0.1)^{10} = \$385.54$$

Therefore, $1000 received in 10 years' time is worth only $385.54 today if the interest rate is 10% p.a. Alternately, $385.54 invested today at 10% p.a. is worth $1000 in 10 years time.

Note that in present worth calculations i is called the **discount rate** because it is used to discount future benefits and costs. As will be discussed later, the value of the discount rate may differ significantly from market interest rates.

Present worth of a uniform series

It is often necessary to find the present worth of a uniform series of constant annual payments. These may represent annual income from sales of a product, or annual maintenance payments on a piece of equipment. Figure 5.2 illustrates a uniform series of payments of $\$A$ per year for n years. By convention, such payments are assumed to occur at the end of each year, so that there is no payment in year 0.

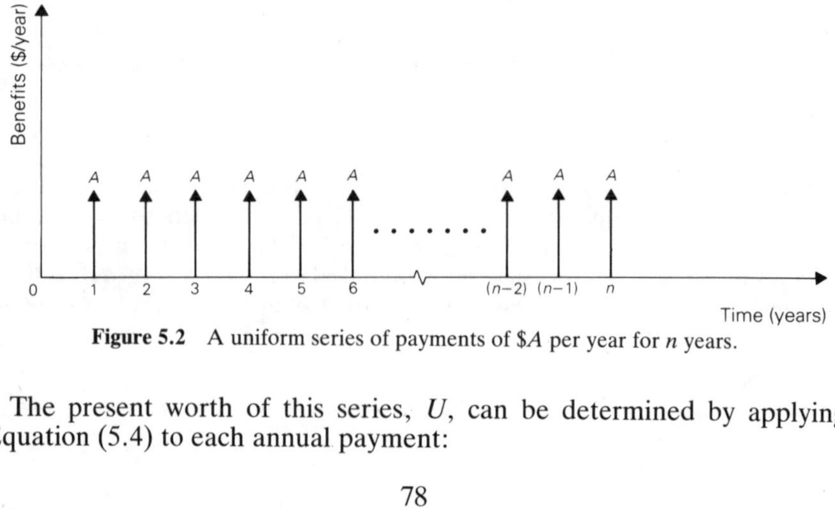

Figure 5.2 A uniform series of payments of $\$A$ per year for n years.

The present worth of this series, U, can be determined by applying Equation (5.4) to each annual payment:

$$U = \frac{A}{(1+i)} + \frac{A}{(1+i)^2} + \frac{A}{(1+i)^3} + \ldots$$

$$+ \frac{A}{(1+i)^n} \tag{5.5}$$

The right-hand side of Equation (5.5) forms a geometric series. A simpler expression for U is determined by first multiplying both sides of Equation (5.5) by $(1+i)$ to give

$$U(1+i) = A + \frac{A}{(1+i)} + \frac{A}{(1+i)^2} + \ldots$$

$$+ \frac{A}{(1+i)^{n-1}} \tag{5.6}$$

Subtracting Equation (5.5) from Equation (5.6) and rearranging, we obtain

$$U = A\left[\frac{1-(1+i)^{-n}}{i}\right] \tag{5.7}$$

This can also be written as

$$U = A\ (P/A,i\%,n) \tag{5.8}$$

Here $(P/A,i\%,n)$ is called the **series present worth factor**.
Transposing Equation (5.7) gives

$$A = U\left[\frac{i}{\{1-(1+i)^{-n}\}}\right] \tag{5.9}$$

$$A = U\ (A/P,i\%,n) \tag{5.10}$$

The term $(A/P,i\%,n)$ is called the **capital recovery factor** and is the reciprocal of the series present worth factor. As with the earlier compound interest factors, these factors are widely tabulated.

In Equations (5.7)–(5.10), $\$U$ is the sum which must be invested now to ensure a series of uniform payments of $\$A$ per year for n years.

Example 5.2 A project yields benefits of \$3M each year for 30 years. If the discount rate is 8% p.a., what is the present value of the benefits?
Solution Using Equations (5.7) and (5.8),

$$(P/A,i\%,n) = \frac{1-(1.08)^{-30}}{0.08} = 11.2578$$

$$U = 3 \times 11.2578 = \$33.77\text{M}$$

Note the effect of discounting future benefits. If the discount rate were zero the present value of benefits would be \$90M instead of \$33.77M.

Another interpretation of this example is that if \$33.77M were invested now in an account paying a constant interest of 8% p.a. it would be possible to withdraw \$3M at the end of each year for 30 years. At the end of this period the balance in the account would be zero.

Effective interest rates

Commercial interest rates are usually quoted as a percentage per year, even if the compounding period is something less than a year (e.g. interest is compounded quarterly, monthly, or daily). In this situation the **effective interest rate** is greater than the quoted **nominal interest rate** and may be determined by application of the compound interest formula.

If the nominal interest rate is $r\%$ per annum and compounding is carried out at the end of m periods per year, then the interest rate applied at the end of each period is $(r/m)\%$. Therefore, if \$$P$ is invested in the account at the start of the year, at the end of one year the compound amount can be determined using Equation (5.3):

$$F = P\{1+(r/m)\}^m \tag{5.11}$$

The effective annual rate, $i\%$, can be determined by applying Equation (5.1) for a single annual period:

$$F = P(1+i) \tag{5.12}$$

Equating the right-hand sides of Equations (5.11) and (5.12) and solving for i gives

$$i = \{1+(r/m)\}^m - 1 \tag{5.13}$$

Example 5.3 A credit agency claims that it charges 18% interest per annum on its loans, but the interest is compounded monthly. What is the effective annual interest rate?

Solution Let i equal the effective annual interest rate. Applying Equation (5.13) with $r = 18\%$ and $m = 12$, we find

$$\begin{aligned} r/m &= 0.18/12 = 0.015 \\ \text{therefore } i &= \{1.015\}^{12} - 1 = 0.1956 \\ \text{or } i &= 19.56\% \end{aligned}$$

The effects of inflation on economic evaluation

Inflation is represented by a rising general level of prices. It is important to note that during an inflationary period not all prices are necessarily rising nor are all prices necessarily rising at the same rate. The effect of inflation is that a dollar next year will buy less than a dollar today.

Changes in the general level of prices may be measured by a **price index**.

This is defined as 'the average price of a mixture of goods and services in a given year' divided by 'the average price of the same mixture of goods and services in a base year' (expressed as a percentage).

A common price index is the **consumer price index** or CPI. The CPI measures changes in the prices of goods and services purchased by moderate-income families. It is based on the prices of a large number of representative items each weighted according to its relative importance in the typical family budget. Movements in the consumer price indexes for a number of countries are shown in Table 5.1.

Table 5.1 Consumer price indexes for various countries.

Country	1970	1982	1986
Australia	100.0	328.1	442.3
Canada	100.0	270.2	324.3
France	100.0	318.7	409.9
West Germany	100.0	183.4	197.7
Italy	100.0	513.7	759.1
Japan	100.0	254.6	272.2
New Zealand	100.0	434.9	660.7
Singapore	100.0	210.0	217.1
United Kingdom	100.0	438.3	532.3
USA	100.0	248.6	283.8

(*Source: Monthly Summary of Statistics*, Australian Bureau of Statistics)

Of particular relevance to engineering projects are the price indexes for construction and building costs. These indexes for a number of North American cities are published in the quarterly *Engineering News Record*. In 1986, the average construction cost index for 20 cities in the USA was 300.8, relative to a value of 100.0 in 1970. When this is compared with a 1986 CPI value of 283.8 for the USA (relative to the same base year) it is apparent that construction costs have risen faster than consumer prices over the last 16 years. In Australia the Australian Bureau of Statistics publishes price indexes for materials used in house building and other building in its *Monthly Summary of Statistics*.

It is most important to distinguish clearly between the concepts of inflation and the time value of money. Even in the absence of inflation, very few people would be prepared to lend money at zero interest. As most people would prefer to consume goods now rather than later, there is clearly a real time value of money. In inflationary times, lenders expect a real rate of return on their money which exceeds the inflation rate. For example, anyone who receives 8% p.a. interest on an investment when the inflation rate is 10% p.a. is clearly losing money.

Before we consider how inflation is taken into account in economic evaluation some definitions are required.

Actual dollars (or **then-current dollars**) are dollars which are current in the particular years that the benefits and costs occur. For example, if a man bought a block of land in 1950 for \$5000 (in then-current dollars) and sold it in 1980 for \$20 000 (in then-current dollars) has he made a profit on the investment? The answer is yes if \$20 000 in 1980 will buy more goods and services than \$5000 in 1950. Otherwise, he has lost on the investment.

Constant-worth dollars are dollars which have the same purchasing power at a defined point in time e.g. 1988 dollars. For example, a new power station is designed to provide 8000 GWh of electrical energy per year under certain operating conditions. The benefits to consumers of this energy are estimated to be \$600M in 1990 (the first year of operation). Thereafter the station will continue to provide 8000 GWh of energy per year for its estimated operating life of 40 years. The annual benefits expressed in actual dollars will continue to increase throughout the life of the station due to inflationary increases in the price of electricity and the prices of goods produced with it. However, when expressed in constant-worth dollars (e.g. 1990 dollars) the annual benefits are likely to be constant over the life of the station as the same amount of energy is produced each year.

The **nominal** interest rate is the rate received on invested money (or paid on borrowed money) when calculated in terms of actual (or inflated) dollars.

The **real** rate of interest on an investment is the rate received net of inflation. For example, if money is invested at a nominal interest rate of 18% p.a. when the annual inflation rate is 10%, the real interest rate (i.e. above inflation) is approximately 8% p.a.

The effect of inflation on the economic evaluation of engineering projects can be taken into account in one of two equivalent ways:

- by using actual dollars and the nominal interest rate throughout
- by using constant-worth dollars and the real rate of interest throughout

The first method is commonly used in evaluating private sector investments in which the future cash flows are estimated in actual dollars and the nominal interest rate is usually known.

The second method is more commonly used for evaluating public sector investments, as it is often easier to estimate future benefits and costs in constant-worth dollars. In this way there is no need to estimate an inflation rate. The discount rate used is the real rate above inflation. As will be discussed in section 5.9, the choice of the discount rate in the public sector is the source of some controversy.

In this book, constant-worth dollars and real interest or discount rates will be used unless otherwise stated.

Relationship between nominal and real interest rates

Consider the annual maintenance costs of a bridge which requires a constant number of person-hours of maintenance each year during its life of n years. (This is a gross simplification, as it would be normal for the amount of maintenance to increase as the bridge deteriorates with age.) When expressed in constant-worth dollars the annual maintenance cost of the bridge is a constant amount (say \$$A$) per year for the life of the bridge, as shown in Figure 5.3. If the anticipated inflation rate for the life of the project is $f\%$ p.a. (compound), then the maintenance costs in actual dollars are shown in Figure 5.4. The nominal rate of interest is assumed to be known and is denoted by i_n. The objective is to find a relationship between the real interest rate, i, the nominal interest rate, i_n, and the inflation rate, f.

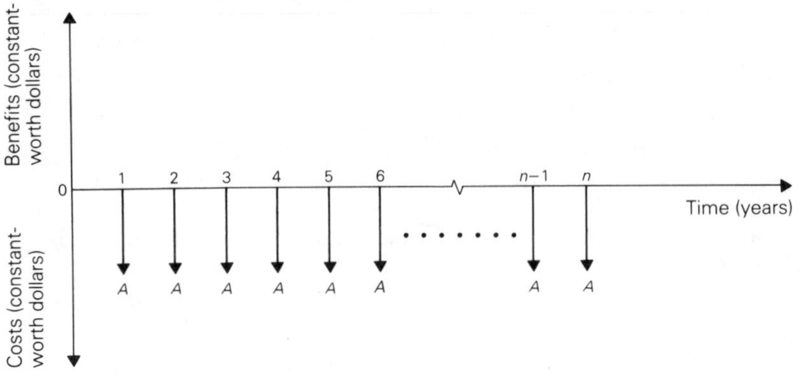

Figure 5.3 Bridge maintenance costs in 'constant-worth' dollars.

The present worth of the series shown in Figure 5.4 is denoted by U. It can be determined by applying Equation (5.4) to each annual payment using actual dollars and the nominal interest rate:

$$U = \frac{A(1+f)}{(1+i_n)} + \frac{A(1+f)^2}{(1+i_n)^2} + \ldots + \frac{A(1+f)^n}{(1+i_n)^n} \tag{5.14}$$

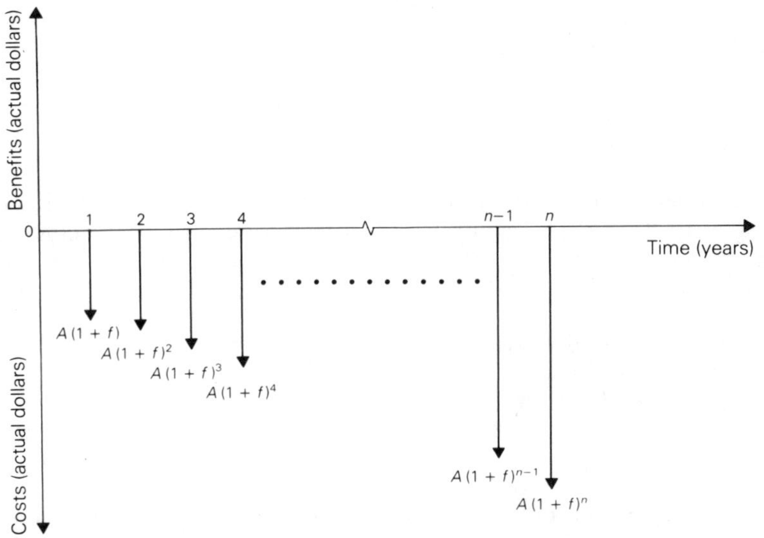

Figure 5.4 Bridge maintenance costs in 'actual' dollars (f = annual inflation rate).

The right-hand side of Equation (5.14) can be simplified, as it is the sum of a geometric series:

$$U = \frac{A(1+f)}{(1+i_n)} \left\{ \frac{1 - (1+f)^n / (1+i_n)^n}{1 - (1+f) / (1+i_n)} \right\} \tag{5.15}$$

Now, the present worth of the series shown in Figure 5.3 is found by using constant-worth dollars and the real interest rate, i. Rewriting Equation (5.5),

$$U = \frac{A}{(1+i)} + \frac{A}{(1+i)^2} + \ldots + \frac{A}{(1+i)^n} \tag{5.16}$$

It should be apparent that the right-hand sides of Equations (5.14) and (5.16) are equal if

$$(1+i) = \frac{(1+i_n)}{(1+f)} \tag{5.17}$$

Equation (5.17) can be used to determine the real rate of interest given a nominal interest rate and the inflation rate. Conversely, the nominal interest rate can be found from the real rate and the inflation rate.

Example 5.4 If money is borrowed at 18% p.a. when the annual inflation rate is 10% p.a., what is the real rate of interest?
Solution Using Equation (5.17) with $i_n = 0.18$ and $f = 0.10$:

$$(1+i) = 1.18/1.10 = 1.0727$$

$$i = 0.0727 = 7.27\% \text{ p.a.}$$

5.4 Evaluation criteria

Various economic criteria have been proposed for use in comparing engineering projects. The following criteria are described briefly and compared:

- payback period
- net present value
- equivalent annual worth
- benefit–cost ratio
- internal rate of return

The payback period

The **payback period** is defined as the time it takes for a project to generate sufficient net benefits to cover its initial cost of construction or implementation.

Example 5.5 A project has an initial capital cost of \$10M, annual benefits of \$3M, and annual operating costs of \$1M. What is its payback period?
Solution Net annual benefits = gross annual benefits minus annual costs.
i.e., Net annual benefits = \$3M − \$1M = \$2M
Therefore, payback period = \$10M/\$2M = 5 years

The payback period criterion does not correctly deal with the time value of benefits and costs. An example of how this criterion gives a false view of the economic merit of a project is illustrated in Figure 5.5. The large benefits which occur after the payback period are not taken into account in its determination. Because of its disadvantages the payback period criterion is not commonly used for evaluating engineering projects. It is not considered further in this book.

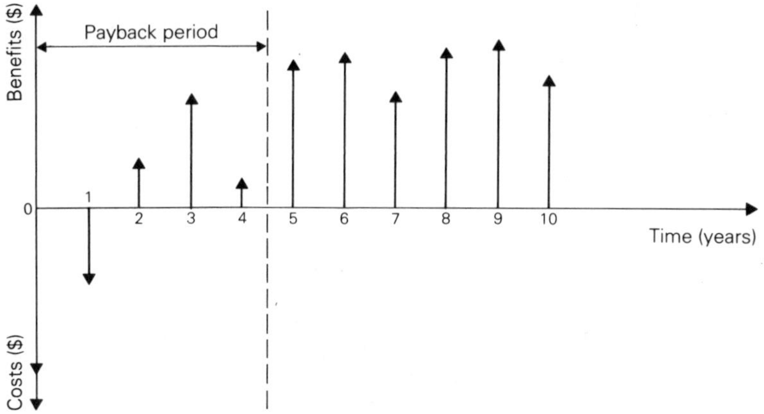

Figure 5.5 The payback period criterion.

Net present value

The **net present value** (NPV) of a project is defined as the present value of all benefits minus the present value of all costs:

$$\mathrm{NPV} = \sum_{t=0}^{n} \frac{B_t}{(1+i)^t} - \sum_{t=0}^{n} \frac{C_t}{(1+i)^t} \tag{5.18}$$

where B_t is the benefit in year t, C_t is the cost in year t, i is the discount rate (% p.a.), and n is the life of the project (years)

NPV involves the use of a discount rate to compare benefits and costs at a single point in time.

Equivalent annual worth

The **equivalent annual worth** (EAW) is found by converting all project benefits into a series of constant annual benefits for n years (where n equals the life of the project). Similarly, all project costs are converted into a series of constant annual costs for n years. The equivalent annual worth is then equal to the equivalent annual benefit minus the equivalent annual cost.

For example, it is desired to calculate the equivalent annual worth for the time stream of benefits and costs shown in Figure 5.6, where n is the life of the project (in years).

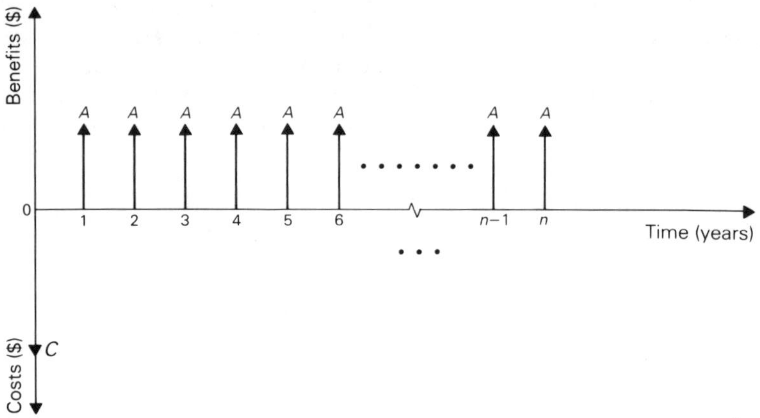

Figure 5.6 A hypothetical time stream of benefits and costs.

In this case, the equivalent annual benefits (EAB) are clearly equal to $\$A$. The equivalent annual cost (EAC) can be determined by converting the initial cost into an annuity for the life of the project using Equation (5.10):

$$\text{EAC} = C(A/P,i\%,n)$$

Therefore, $\text{EAW} = \text{EAB} - \text{EAC} = A - C(A/P,i\%,n)$
For comparison, $\text{NPV} = A(P/A,i\%,n) - C$

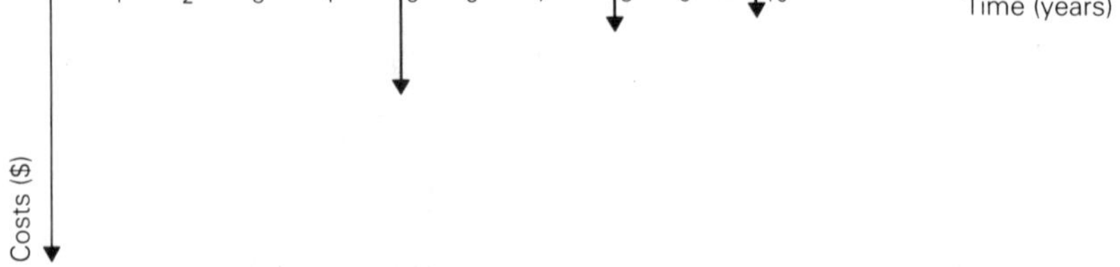

Figure 5.7 An irregular time stream of benefits and costs.

In the above example it is just as easy to calculate the EAW as the NPV. However, this is not always the case. For example, to determine the EAW of the general time stream of benefits and costs illustrated in Figure 5.7 it is first necessary to convert all payments to their value at a common point in time. For convenience this will usually be taken as year 0, although any other time can be used. Equation (5.18) can be used for this purpose. Then Equation (5.10) can be used to convert the single lump sum in year 0 into a uniform payment throughout the life of the project.

Benefit–cost ratio

The **benefit–cost (B/C) ratio** is widely used for evaluating projects in the public sector. As normally defined, it is the present value of all benefits divided by the present value of all costs:

$$\mathrm{B/C} = \frac{\displaystyle\sum_{t=0}^{n} \frac{B_t}{(1+i)^t}}{\displaystyle\sum_{t=0}^{n} \frac{C_t}{(1+i)^t}} \tag{5.19}$$

An alternative definition involves the inclusion of only capital costs in the denominator of Equation (5.19). In this case, the present value of annual operating, maintenance, and repair costs are subtracted from benefits in the numerator.

Internal rate of return

The above economic criteria (with the exception of the payback period) involve the use of a discount rate to convert all future benefits and costs into present value. The internal rate of return (which is also called the **yield**) is a different concept in that it is the implied interest rate of the investment. In other words, the internal rate of return is the discount rate at which the present value of benefits just equals the present value of costs for the life of the project.

The **internal rate of return** (IRR) of a project, r, may be found by solving the following equation:

$$\sum_{t=0}^{n} \frac{B_t}{(1+r)^t} = \sum_{t=0}^{n} \frac{C_t}{(1+r)^t} \tag{5.20}$$

In general this will involve the solution of a polynomial equation in order to determine r.

5.5 A comparison of the economic criteria

Types of decisions requiring an economic evaluation criterion

The following decision-making situations may be identified:

- Go or no-go decisions. For example, the decision of whether or not to build a new airport or whether or not to build a long and expensive railway line.
- The choice of a single project from a list of mutually exclusive projects (which may include the 'no build' alternative). For example, the choice between a bridge and a tunnel for crossing a waterway.
- The choice of a number of projects from a list of feasible projects when the total funds available are limited (this is often called the **capital budgeting problem**). An example is the problem faced by a state water authority when deciding how to allocate its annual capital works budget among the many feasible projects available.

GO OR NO-GO DECISIONS

In deciding whether or not an individual project is justified on economic grounds, any one of the following criteria may be applied:

- NPV > 0
- EAW > 0
- B/C > 1
- IRR > i

Only one of these criteria need be examined, as all will lead to the same decision for an individual project: i.e., if the present value of the benefits exceeds the present value of the costs, then the NPV will be positive, the EAW will be positive, the B/C ratio will exceed 1 and the internal rate of return will exceed the discount rate. Therefore only one criterion needs to be used.

THE CHOICE OF A SINGLE PROJECT FROM A LIST OF MUTUALLY EXCLUSIVE PROJECTS

In this case, unlike the go or no-go situation, the final decision may depend on the economic criterion used. This is illustrated by the following example:

A cartage contractor is considering buying a fleet of new trucks. Two basic models are available, each of which costs $100 000. Economic data relating to the two trucks are given in Table 5.2.

Table 5.2 Economic parameters for the two trucks.

Truck model	Cost ($)	Annual benefits ($)	Scrap value ($)	Life (years)
Itzumi	100 000	20 000	10 000	10
Ward	100 000	15 000	10 000	20

The annual benefits of each truck are estimates of the annual revenue less the annual operating costs. The life of each model is the time after which it is either obsolete or too expensive to keep running. The scrap value is the estimated resale value at the end of its life. The appropriate discount rate for the contractor is 10% p.a.

Which truck should the contractor buy? Assume that he asks his daughter (a graduate in civil engineering) to carry out an economic analysis. She is unsure of which economic criterion to use, but produces the following figures:

For the Itzumi truck

$$\begin{aligned}\text{Present value of benefits} &= 20\,000\,(P/A,10\%,10) + 10\,000\,(P/F,10\%,10)\\ &= 20\,000 \times 6.1446 + 10\,000 \times 0.3855\\ &= \$126\,750\end{aligned}$$

$$\begin{aligned}&\text{Present value of costs} && = \$100\,000\\ &\text{NPV} = \$126\,750 - \$100\,000 && = \$26\,750\\ &\text{EAW} = \$26750\,(A/P,10\%,10) && = 26750 \times 0.1627 = \$4353/\text{year}\\ &\text{B/C} = 126\,750\,/\,100\,000 = 1.267\end{aligned}$$

The internal rate of return can be found by solving the following equation for r:

$$20\,000\,\{1 - (1+r)^{-10}\}\,/\,r + 10\,000\,(1+r)^{-10} - 100\,000 = 0$$

Solving by trial and error gives $r = 0.157$, i.e. the internal rate of return is 15.7%.

Similar calculations can be carried out for the Ward truck. A comparison of the economic performance of each truck is given in Table 5.3.

Table 5.3 Economic comparison of the two trucks.

Criterion	Truck		Choice
	Itzumi	Ward	
NPV ($)	26 750	29 190	Ward
EAW ($/year)	4353	3429	Itzumi
B/C	1.267	1.291	Ward
IRR (%)	15.7	14.0	Itzumi

Here we have the confusing result that two of the economic criteria favour the Itzumi truck while the other two favour the Ward truck. Clearly the decision reached depends on the criterion used.

Another complicating factor in the above analysis is that the alternatives have different lives. The comparison in Table 5.3 is not totally valid for this reason. As a general rule we should always compare alternatives over the same planning horizon. In the above example this can be achieved by assuming that an Itzumi truck will be replaced by another Itzumi truck at the end of its life. Therefore the time stream of benefits and costs for this alternative is shown in Figure 5.8.

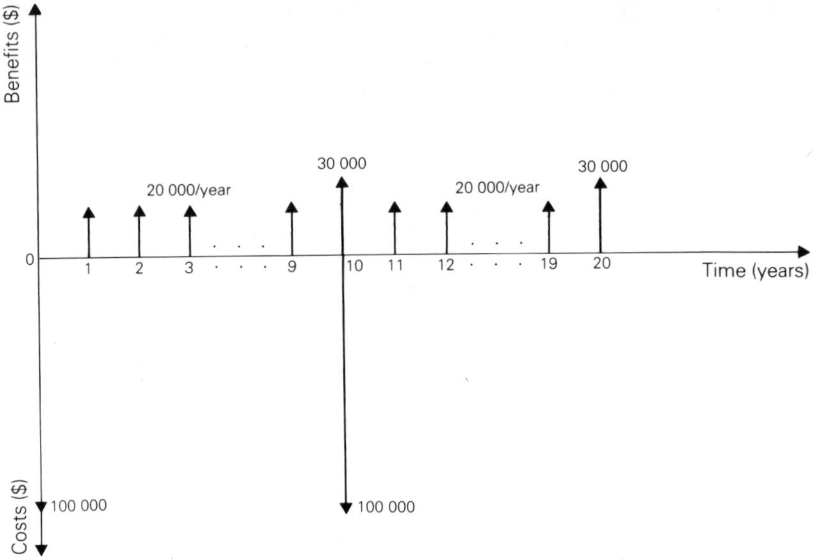

Figure 5.8 The time stream of benefits and costs achieved by the purchase of Itzumi trucks.

The present value of the benefits and costs of purchasing Itzumi trucks can now be determined as:

$$\text{PV of benefits} = \$126\,750 + \$126\,750\,(P/F,10\%,10) = \$175\,620$$

$$\text{PV of costs} = \$100\,000 + \$100\,000\,(P/F,10\%,10) = \$138\,550$$

Using these figures the revised values of the criteria are given in Table 5.4.

Table 5.4 Revised economic comparison of the trucks.

Criterion	Truck		Choice
	Itzumi(2)	Ward	
NPV ($)	37 070	29 190	Itzumi
EAW ($)	4353	3429	Itzumi
B/C	1.267	1.291	Ward
IRR (%)	15.7	14.0	Itzumi

Note that the assumption of a repeated purchase of the Itzumi truck affected its NPV but none of the other economic criteria.

As the choice of project may depend on the criterion used, we need to examine the advantages and disadvantages of each. This is an issue which will be addressed in the section later in this chapter headed 'Advantages and limitations of each criterion'. In general, it is recommended that the NPV criterion is used for this type of decision.

LEAST COSTS DECISIONS

A variation on the above decision-making situation is when one design must be chosen from among a number of alternatives, each of which substantially performs the same function. For example, the choice between two designs for a road bridge, one of which has prestressed concrete girders and the other steel girders. In both cases, the girders act compositely with a reinforced concrete deck. The benefits of the bridge may be assumed to be the same for both designs and may therefore be ignored in the decision. In this case the NPV criterion is equivalent to finding the design which minimizes the present value of costs (assuming that the present value of benefits exceeds the present value of costs for the chosen design). Alternatively the choice may be made by minimizing the equivalent annual cost. The B/C and IRR criteria cannot be used in this situation.

THE CAPITAL BUDGETING PROBLEM

In this case the usual procedure when using the B/C ratio or IRR is to rank the projects from best to worst in terms of the criterion. Projects are then chosen from the top of the list down, until the total capital budget is exhausted. When using NPV or EAW the procedure is a little more complicated in that a set of projects must be chosen which maximize the appropriate criterion, subject to constraints on the available capital budget.

As with the previous decision problem, each of the economic criteria could identify a different set of projects to be undertaken.

Example 5.6 A list of possible road projects is given in Table 5.5. Which projects would be selected using B/C ratio and NPV as the economic criterion, if the available capital budget is $35M?

Table 5.5 Economic comparison of a set of road projects. PVB is the present value of benefits, and PVC is the present value of costs (assumed to be all capital costs in this case).

Project	PVB($M)	PVC($M)	NPV($M)	B/C
A	40	15	25	2.67
B	18	10	8	1.80
C	18	12	6	1.50
D	50	20	30	2.50
E	28	10	18	2.80
F	26	20	6	1.30

Solution Using B/C ratio as the selection criterion the projects are first ranked in order of this measure as shown in Table 5.6. Projects are then selected from the top of the list down until the capital budget is exhausted. Projects E and A would be selected first at a total capital cost of $25M. Project D cannot be selected as its capital cost of $20M exceeds the $10M remaining in the capital budget. The next best project is B whose capital cost of $10M can be accommodated within the budget.

Table 5.6 The road projects in order of decreasing B/C ratio.

Project	PVB($M)	PVC($M)	NPV($M)	B/C
E	28	10	18	2.80
A	40	15	25	2.67
D	50	20	30	2.50
B	18	10	8	1.80
C	18	12	6	1.50
F	26	20	6	1.30

Therefore using B/C ratio the selected projects are: E, A, and B.

Using NPV we need to set up a simple optimization model. Firstly, define a set of variables X_i (i = A,B,C,D,E,F) such that $X_i = 1$ if project i is chosen and = 0 otherwise.

Then the set of projects which maximize NPV can be found by solving the following integer linear programming problem:

$$\text{Maximize } Z = 25X_A + 8X_B + 6X_C + 30X_D + 18X_E + 6X_F$$
$$\text{subject to } 15X_A + 10X_B + 12X_C + 20X_D + 10X_E + 20X_F \leqslant 35$$
$$X_i = 0,1 \ (i = \text{A,B,C,D,E,F})$$

The solution of integer linear programming problems is beyond the scope of this book. The use of linear programming for continuous variables is dealt with in Chapter 7.

The solution to this sample problem can be found by inspection. It is $X_A = X_D = 1$; all other $X_i = 0$; i.e. choose projects A and D at a total cost of $35M and NPV = $55M. The solution using B/C ratio gave a NPV of $51M and therefore involved a loss in potential net benefits of $4M or 7% when compared to the optimum.

This example demonstrates that NPV is a better criterion than B/C for capital budgeting decisions. Similar examples can be produced to show that IRR is also inferior to NPV in capital budgeting. EAW will produce identical results to NPV if all projects have the same life.

Advantages and limitations of each criterion

From the previous section it should be clear that each economic criterion may lead to a different choice of project(s) when choosing one or more from a list of feasible alternatives. It is important therefore to consider the advantages and limitations of each criterion in turn.

THE INTERNAL RATE OF RETURN

This criterion has the advantage of not requiring a prescribed value of discount rate for its determination. As the determination of a discount rate may be the source of considerable controversy, particularly in the public sector, this is an advantage in project selection. However, this advantage is outweighed by the disadvantages of the criterion, namely:

- to determine the IRR of any project involves the solution of a polynomial equation
- sometimes more than one value of IRR may be obtained for a particular project

The first point was illustrated in Section 5.5 where the IRR of two projects was determined. For a complicated time stream of benefits and costs, the value can only be determined by trial and error.

The second point can be illustrated with an example.

Example 5.7 Figure 5.9 shows the time stream of expected revenue and costs for a proposed open-cut mining development. There is an initial cost to establish the mine and associated works and a high cost at the end of the project due to clean-up and re-vegetation activities. What is the IRR for the project?

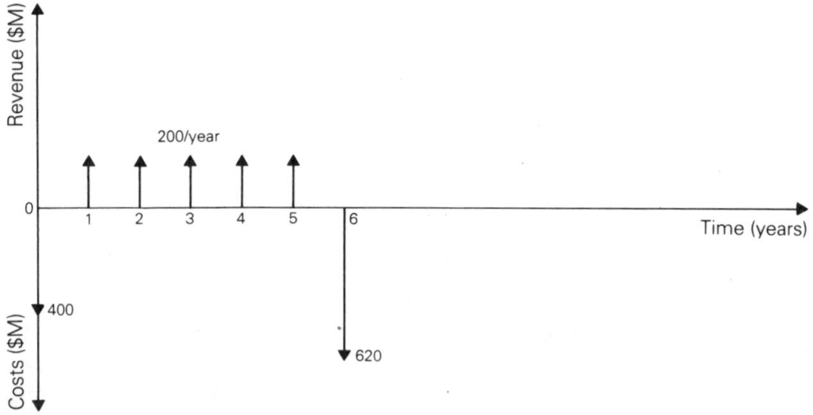

Figure 5.9 The time stream of benefits and costs for a proposed open-cut mining development.

Solution Let i = the discount rate for the project. Then,

$$\text{NPV (\$M)} = \frac{200\,\{1-(1+i)^{-5}\}}{i} - \frac{620}{(1+i)^6} - 400 \qquad (5.21)$$

Figure 5.10 shows a plot of NPV as a function of i. It can be seen that there are two values of i for which the NPV is zero, namely 3.8% and 16%. Both of these values may be interpreted as the internal rate of return of the project. If the project were being ranked in a capital budgeting exercise it could receive a high or low place depending on the value of IRR chosen. Although this is a somewhat unusual case it does demonstrate one difficulty with using the IRR criterion, i.e. it is the solution of a polynomial equation for which more than one root may exist.

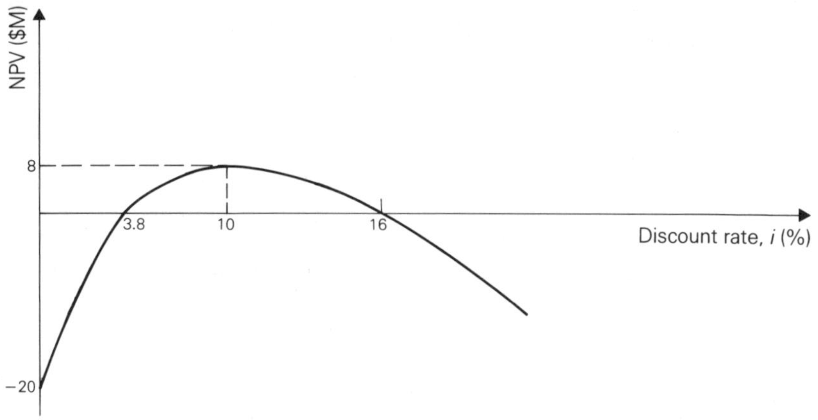

Figure 5.10 NPV as a function of discount rate for the proposed open-cut mining development.

EQUIVALENT ANNUAL WORTH

To determine the EAW of a general time series of benefits and costs it is necessary first to compute the NPV of the series and convert the NPV into a uniform annual payment throughout the life of the project (using Equation (5.10)). Considering that EAW gives the same ranking as NPV for projects with equal lives, a strong case can be made for using NPV directly.

BENEFIT–COST RATIO

The benefit–cost ratio (B/C) has been used extensively in the USA, UK, Australia and elsewhere for comparison of projects of various sizes. It has some degree of appeal when used in the context of allocating scarce capital funds, i.e. capital budgeting. However, there is some ambiguity in its definition for projects with high recurrent costs (e.g. operating, maintenance, and repair costs). The basic issue here is whether the present value of recurrent costs should be treated as negative benefits and deducted from the numerator of the B/C ratio or treated as costs and added to the denominator. A different value of B/C ratio will result in each case.

For example, consider the economic comparison of two projects with the relevant economic data summarized in Table 5.7.

Table 5.7 Summary economic data for two projects. PV is present value, OMR are operating, maintenance, and repair costs.

Economic parameter	Project 1	Project 2
PV of benefits, B ($M)	1.5	4.0
Construction cost, C ($M)	1.0	1.0
PV of OMR costs, K ($M)	0.0	2.0
B/C Ratio No. 1 = $B/(C+K)$	1.50	1.33
B/C Ratio No. 2 = $(B-K)/C$	1.50	2.00
NPV = $B-K-C$	0.5	1.0

For B/C Ratio No. 1 in Table 5.7 the present values of all costs are included in the denominator. For B/C Ratio No. 2 the present values of recurrent costs are subtracted from the present value of benefits in the numerator. The rationale for this latter definition is that in many cases it is really the capital costs which are limited; the recurrent costs may be met out of the annual benefits.

If the projects are compared using the first definition of B/C ratio, Project 1 is preferable. Conversely, if the second definition of B/C ratio is used, Project 2 is preferable. On the other hand, NPV always gives an unambiguous ranking of projects as it does not matter whether OMR costs are first subtracted from the benefits or added to the construction costs. In this case, the NPV criterion favours Project 2.

It might be argued that this is not a major problem; why not just define the B/C ratio in one way or the other and proceed? However, in many situations the project benefits are composed entirely of savings in cost. For example, the benefits of most road projects consist of savings in travel time, savings in operating cost, and savings in cost due to a reduction in accidents. In such cases the B/C ratio obtained depends on the arbitrary definitions of benefits and costs, whereas NPV is unambiguous.

NET PRESENT VALUE

Most economists advocate NPV as the most appropriate criterion for comparing projects. It measures the net gain to society (or the company) of the proposed project. It can be used in any of the three decision-making contexts discussed above. In the case of a go or no-go decision the choice is independent of the criterion used. In choosing a single project from a list of mutually exclusive projects, NPV appears to favour large projects.

For example, an engineering firm is considering the purchase of a microcomputer to meet some of its computing needs. The firm currently pays for the use of a mainframe computer on a time-sharing basis. Two models of microcomputer are being considered. The relevant economic data are given in Table 5.8. The benefits represent the estimated savings in computing cost due to reduced usage of the mainframe computer. The present value of benefits for each microcomputer is computed using Equation (5.7).

Table 5.8 An economic comparison of two microcomputers. (Discount rate = 12%.)

	Micro A	Micro B
Purchase price ($)	10 000	20 000
Annual benefits ($)	4000	8000
Life (years)	5	5
PV of benefits ($)	18 914	37 828
NPV ($)	8914	17 828
EAW ($/year)	1226	2452
B/C	1.89	1.89
IRR (%)	28.65	28.65

From Table 5.8 it can be seen that microcomputer B has the same life as microcomputer A, but twice the cost and twice the benefits. Its NPV and EAW are therefore twice those of microcomputer A while its B/C ratio and IRR are the same. The tendency for NPV (and EAW) to favour large projects is apparent. If there is no constraint on the amount of money that can be spent, the increased NPV of microcomputer B is a net gain to the firm and therefore it is the better option. If there are limited funds available for investment due to other equipment needs, the problem should be treated as a capital budgeting exercise as discussed above.

A related problem is the choice of the optimum size of an individual project.

For example, consider a dam on a river which is to supply water to an

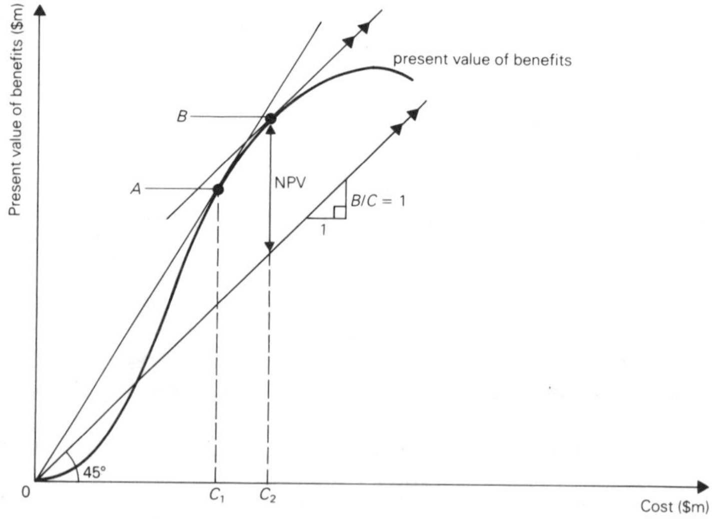

Figure 5.11 The selection of optimum project size (adapted from deNeufville and Stafford, 1971).

adjacent irrigation area. We assume that the only cost associated with the dam is its initial construction cost. The PVB depends on the size of the dam and hence on its initial cost. A typical pattern of PVB versus initial cost is shown in Figure 5.11. Clearly, a very small dam would not supply much water and therefore the PVB would be small. As the size of the dam is increased an efficient size is reached as indicated by a steep rise in the PVB curve. For very large dams, the PVB curve tends to flatten out as the dam is likely to be less than full for most of the time.

The optimum dam size (or cost) is that which maximizes NPV. This will occur at the point where the marginal benefit of an extra dollar of cost equals unity, i.e. where the slope of the PVB curve equals 1. In Figure 5.11 this occurs at point B.

The size of dam which maximizes the B/C ratio is identified by the point where a straight line drawn from the origin is tangent to the highest possible point on the curve (i.e. point A in Fig. 5.11). The B/C ratio for the design corresponding to point A is given by the slope of the line to the origin. As the slope of the PVB curve at point A is greater than unity, it should be clear that if the dam cost is increased by \$1, the increase in benefits will be greater than \$1. Thus, by increasing the dam cost (and hence size) from C_1 to C_2 (i.e. by maximizing NPV instead of B/C ratio) a net gain in community benefits is achieved.

Figure 5.11 illustrates that by selecting the project size so as to maximize B/C ratio the result will generally be a project which is smaller than the optimum.

With regard to capital budgeting decisions, it was shown above that the use of NPV may lead to higher net benefits when compared with the other criteria. Its use is therefore favoured in this situation also.

5.6 Economic benefits

One important issue which needs to be addressed in economic evaluation is the estimation of the benefits and costs of any particular project.

In order to understand the basis for this estimation some concepts from microeconomic theory must be utilized. The theory of price determination in perfectly competitive markets is presented in the Appendix to this chapter.

A fundamental distinction should be drawn between the estimation of benefits in the private and public sectors.

In the private sector, the benefits (or revenue) of a particular project may be identified as **cash flows** into the firm. Benefits may be determined by the additional output of goods and services multiplied by the market prices for these commodities. In a competitive market situation each individual firm is obliged to sell its output at the ruling market price due to pressure from its competitors. The output of any one firm is so small that it has a negligible impact on the market price. Furthermore, the market price is a measure of the willingness-to-pay of the marginal consumer of that product. Therefore, the market price is a measure of the marginal benefit to society of providing additional units of the commodity.

In the public sector, the basic economic objective is to maximize benefits minus costs for society as a whole. In many cases benefits (and costs) will not equal the cash flows which result from the project.

The theoretically correct measure of public sector benefits is the **willingness-to-pay** by consumers for the outputs of the project. The amount which a consumer is willing to pay for a commodity is a measure of the value to that consumer of the commodity. This concept is illustrated in Figure 5.12 for the provision of a new good (e.g. video recorders) to a community. It is assumed that there is a maximum price, P_1, that any one person is willing to pay for the new good. The market price, as determined by the intersection of the demand and supply curves for the good, is P_0. However, as indicated by the demand curve, one consumer is willing to pay $\$P_1$ for the good. Thus $\$P_1$ is a measure of the value of the good to that consumer. Someone is willing to pay $\$P_2$ ($< \$P_1$) for the second unit and so on. Finally, someone is only willing to pay $\$P_0$ for the good. Anyone not willing to pay $\$P_0$ will, of course, not purchase the good at the current market price. Therefore, the total willingness-to-pay for the good is represented by the total shaded areas (A+C) in Figure 5.12. The amount actually paid by the consumers will equal the **revenue** P_0Q_0 (area C). The difference between the willingness-to-pay and the revenue is called the **consumers' surplus** (i.e. the area A). It represents the *net* benefits to the consumers of the good. The revenue is a benefit to the producers.

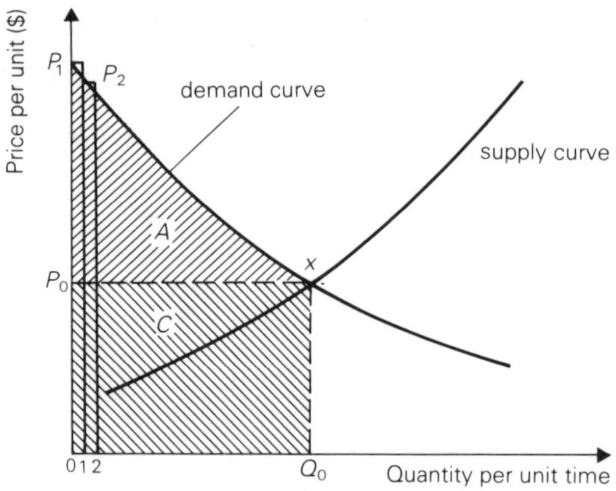

Figure 5.12 The willingness-to-pay concept.

In many cases the benefits of a public project can be assessed by the increase in consumers' surplus resulting from the project. For example, a state highway authority proposes to duplicate an existing two-lane highway which is currently heavily congested. The benefits of this project can be assessed using the increase in consumers' surplus. Figure 5.13 shows the demand–supply situation for the highway. The price per trip is represented by the sum of the operating cost plus the value of travel time per trip. The supply curve for the existing highway represents the price per trip as a function of the two-way volume.

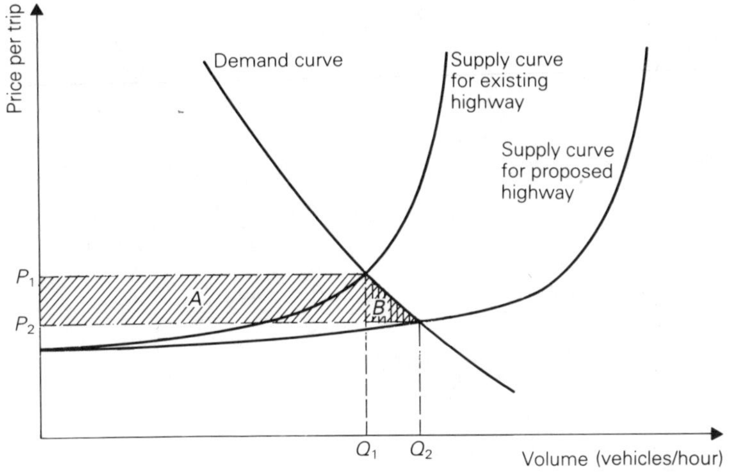

Figure 5.13 Increase in consumers' surplus of a duplicated highway.

The equilibrium volume on the existing highway is Q_1 at a cost per trip of P_1. The new equilibrium volume on the duplicated highway will be Q_2 at a cost per trip of P_2. Note that an increase in volume is expected because a number of motorists now find it attractive to travel on the highway at the reduced price of P_2 per trip.

The benefit to all motorists is equal to the increase in consumers' surplus which is the shaded area in Figure 5.13. This consists of two parts:

- a reduction in cost to the existing users of the road (i.e. the rectangular area A)
- the consumers' surplus to new users (i.e. the approximately triangular area B)

If the demand curve is approximately linear, the benefits are given by:

$$\text{Increase in consumers' surplus} = (P_1 - P_2)\ \left(\frac{Q_1+Q_2}{2}\right) \qquad (5.22)$$

Note that the concept of revenue is not relevant in this problem as the price is not paid to a producer but represents a resource cost to society.

As a second example, the demand–supply situation for the provision of water to a city by a water authority is shown in Figure 5.14. The authority is considering constructing a new dam which would cause the supply curve to move to the right as shown. Assume that the authority sets the price of water equal to that which would apply in a competitive market situation. Under the existing situation Q_1 units of water are supplied per year at a price of P_1. With the new dam in place, the price would fall to P_2 and the volume supplied would increase to Q_2.

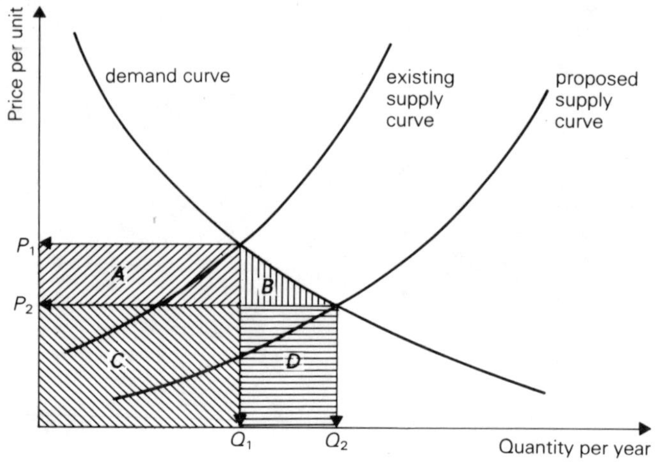

Figure 5.14 Benefits of an increase in urban water supply under market pricing.

The benefits of the new dam will consist of the following:

- the benefits to the consumers as represented by the increase in consumers' surplus. This is area (A+B) in Figure 5.14
- the increase in revenue to the water authority. This is given by $(P_2Q_2-P_1Q_1)$ in Figure 5.14 or area (C+D) minus area (A+C).

The total benefits will be the sum of these, i.e.

$$\begin{aligned}\text{total benefits} &= \text{increase in consumers' surplus} \\ &\quad \textit{plus} \text{ increase in revenue} \\ &= (A+B) + [(C+D) - (A+C)] \\ &= (B+D)\end{aligned}$$

As shown in Figure 5.14 this is simply the increase in willingness-to-pay of the consumers. Assuming a linear demand curve it is given by

$$\text{total benefits} = (Q_2-Q_1)\left(\frac{P_1+P_2}{2}\right) \tag{5.23}$$

If the increase in Q is small, $P_1 \simeq P_2$ and Equation (5.23) becomes

$$\text{total benefits} = (Q_2 - Q_1) \times P_1 \tag{5.24}$$

i.e.

$$\text{total benefits} = \text{increase in output} \times \text{market price}$$

Note that in the case of a large public authority (or a private monopoly) an increase in supply could be so large as to cause a price change.

Now suppose the water authority does not charge the market price for water, but a price of P_3 as set by the government (see Figure 5.15). If the initial quantity supplied is Q_1 the authority would have to ration the water between consumers as the total demand Q_3, at a price P_3, exceeds the available supply Q_1. This is common practice in the supply of water for irrigation purposes in which each irrigator has an annual allowance which cannot be exceeded under normal circumstances.

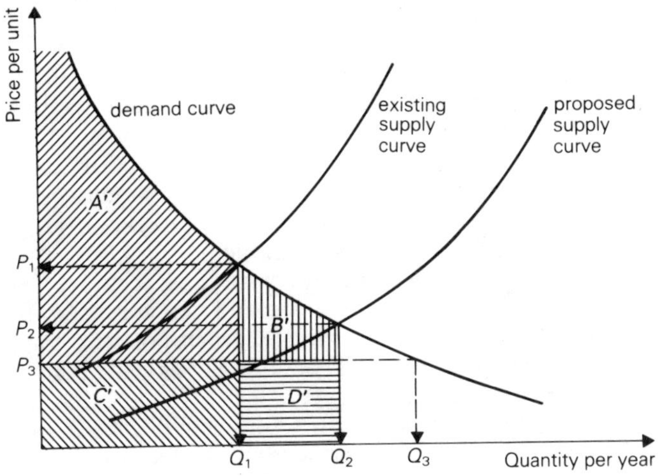

Figure 5.15 Benefits of an increase in urban water supply under non-market pricing.

With the existing supply curve and a rationed quantity Q_1 the total willingness-to-pay is given by areas (A′+C′). This is composed of the authority's revenue (area C′) and the consumers' surplus (area A′). The quantity $Q_1(P_1-P_3)$ is the difference between what could be charged for the water under a competitive market system and what is actually charged. It represents a **transfer payment** from the authority to the consumers as a result of the government pricing policy.

If the supply curve shifts as a result of building a new dam and quantity Q_2 is supplied at price P_3, the increase in benefits will be the increase in consumers' surplus plus the increase in revenue, i.e. area (B′1+D′) in figure 5.15. This is identical to the increase in willingness-to-pay given by (B+D) in Figure 5.14. Therefore the government pricing policy does not affect the total benefits of the scheme, only the way in which these are distributed between consumers and the water authority.

5.7 Economic costs

The concept of cost is a subtle one which may cause confusion in some engineering studies. Essentially, the same concept of cost is applicable to public sector and private sector projects, although, as with benefits, the question of cost *to whom* must be addressed. For the private sector, only costs incurred by the firm are of relevance, whereas for public sector projects the total cost to all members of society must be evaluated. The correct concept of cost is the **incremental opportunity cost** of resources used in the project.

The **opportunity cost** of a resource is its value when used in the best available alternative use. This is a measure of the value which must be forgone by using the resource in the project under consideration. The opportunity cost may be quite different from what was actually paid for the resource.

For example, a company has a vacant block of land which is valued at \$500 000 on the open market. The company paid only \$100 000 for the land some 10 years ago. If the company is considering building an office block on the land, the opportunity cost of the land is at least \$500 000 because the land could be sold at this price. This is the cost which should be used in an economic evaluation of the proposed office block. The opportunity cost may be higher if a better alternative use is available.

As a further example of this concept, consider a contractor who has a crane which costs \$500 per day to operate and which he can hire out for \$1000 per day. The opportunity cost to the contractor of using the crane on a particular construction project is \$1000 per day as this is the benefit which he must forgo in order to use the crane.

The opportunity cost concept has particular significance in public sector evaluation. If, for example, a project in the public sector employs labour which was previously unemployed, the opportunity cost of doing so is considerably less than the actual wage rate. As the alternative use of the labour is, in fact, unemployment, the opportunity cost is the value which the workers place on their additional recreation time when unemployed. The question of unemployment benefits does not enter the evaluation as these are a **transfer payment** between one section of the community and another, i.e. the benefit of this payment to the unemployed equals the cost to the taxpayers, so its net effect on the total community is zero.

The **incremental** or **marginal** costs of a project must be carefully assessed. In project evaluation the marginal costs are the additional costs of undertaking the project compared with not undertaking it.

There is sometimes a tendency to confuse marginal cost with average cost. The distinction between the two is illustrated by the following example.

An engineer purchases a new car for \$20 000. She uses the vehicle to commute to work and for recreational travel on weekends. The estimated annual costs for the vehicle are given below:

Standing costs

	Loss of interest on capital	$ 3000
	Depreciation	$ 4500
	Registration	$ 300
	Insurance	$ 400
Running costs		
	Maintenance and repairs	$ 400
	Petrol and oil	$ 1200
	Tyres	$ 200
Total annual costs		$10 000

If she travels 20 000 km per year, the average cost of travel is

$$\frac{10\,000}{20\,000} = \$0.50/\text{km}$$

Now if her employer asks her to use her own vehicle to make a single trip of 20 km to a construction site, what is the cost of this trip? One possible answer would be to use the average cost of $0.50/km and thereby estimate a trip cost of $10. However, the extra trip is incremental to her private travel and therefore the incremental or marginal cost should be used. The registration and insurance costs and loss of interest on capital are unaffected by the extra trip and hence are irrelevant in this situation. Likewise depreciation is more strongly related to the ageing of the vehicle rather than distance travelled so it does not enter the calculation. The only incremental costs are petrol and oil, tyres, and maintenance and repairs (assuming these depend on distance travelled). Therefore, the marginal cost for a short trip is given by:

$$\frac{400 + 1200 + 200}{20\,000} = \$0.09/\text{km}$$

This gives a cost of $1.80 for the 20 km trip. If the engineer is reimbursed at the average cost of $0.50/km, she makes a nice profit on the trip.

Now consider a different situation in which the engineer uses her car for 15 000 km of business travel and 5000 km of personal travel per year. It could be argued in this case that the vehicle is primarily for business purposes and the private travel is the marginal component. It would then be reasonable for the company to reimburse all standing costs of the vehicle as well as the running costs associated with the business travel. Alternatively the company may provide a vehicle and ask the engineer to pay the running costs associated with private travel.

A further distinction needs to be made between **sunk** and **recoverable** costs. Sunk costs are costs which have been incurred in the past and are no longer recoverable. As such they are irrelevant to decisions about future actions.

Example 5.8 A firm has a photocopying machine which originally cost \$12 000 to purchase and now costs \$2500 per year to operate and maintain. It can be sold for \$5000. An equivalent new photocopying machine costs \$15 000 but it is estimated that it costs only \$1500 per year to run because of an attractive maintenance contract being offered by the manufacturer. What are the benefits and costs which should be considered in deciding whether or not to purchase the new machine? A time-frame of 5 years should be used in analysing the decision. At the end of this time the new machine is expected to be worth \$4000 and the old machine nothing.

Solution We need to consider the incremental benefits and costs of the new machine relative to the old. These are shown in Figure 5.16.

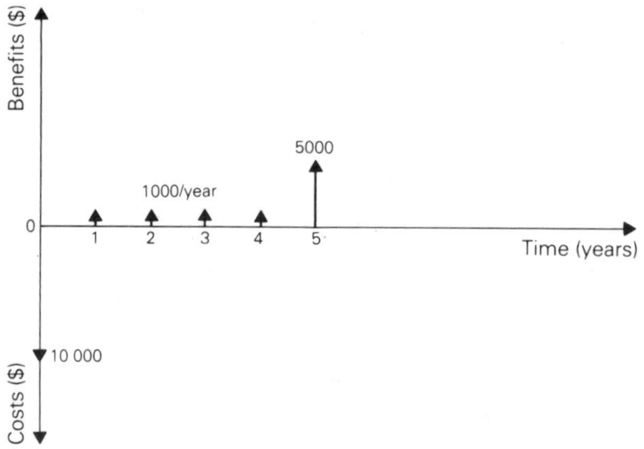

Figure 5.16 Benefits and costs of purchasing a new photocopying machine.

The incremental costs of purchasing the new machine is \$15 000 minus the \$5000 sale value of the old machine. The incremental benefit is the \$1000 per year savings in operating cost plus the \$4000 value of the new machine at the end of 5 years. Note that the original purchase price of the old machine (i.e. the \$12 000) is a sunk cost and does not enter the decision. Clearly the decision would be no different if the old machine had cost \$200 000 or had been obtained for nothing.

5.8 Cost estimation

A common method for estimating the costs of engineering projects involves the use of unit costs. A **unit cost** is the average cost per unit of material or resource. Examples include a unit cost per tonne of steel and a unit cost per cubic metre of excavation.

Suppose we wish to estimate the structural cost of a proposed multistorey building of reinforced concrete frame construction. In the first place a preliminary design should be undertaken to estimate the spacing and sizes of columns, beams, and other structural members. From this a table of quantities can be drawn up such as that shown in Table 5.9.

Table 5.9 Estimated quantities for a reinforced concrete frame building.

Volume of concrete in columns	20 m^3
Volume of concrete in beams	24 m^3
Volume of concrete in slabs	290 m^3
Volume of concrete in foundations	40 m^3
Area of formwork	1400 m^2
Mass of reinforcing steel	40 tonnes
Volume of excavation	480 m^3

A set of unit costs should be available from other similar jobs. These are usually 'in-place' costs including all labour and overheads. A typical set of unit costs is given in Table 5.10.

Table 5.10 Unit costs for a reinforced concrete frame building.

Concrete in columns	\$162/m^3
Concrete in beams	\$150/m^3
Concrete in slabs	\$130/m^3
Concrete in foundations	\$122/m^3
Formwork	\$60/m^2
Reinforcing steel	\$1250/tonne
Excavation	\$12/m^3

By multiplying the quantities by the unit costs an estimated \$189 000 is obtained for the structural cost of the building.

5.9 Selection of project life and discount rate

The choice of project life and discount rate may have an important influence on the economic viability of a particular project. Many engineering projects (e.g. roads, water supply, electricity supply) involve high initial costs and benefits which grow steadily with time over many years. The NPV of such projects falls with increasing values of the discount rate, and may become negative for high values.

For example, a proposed new highway will cost \$10M to construct. The benefits due to savings in operating costs and travel time are estimated to be \$1.2M in the first year of operation but will grow at 2% p.a. (compound) due to growth in the volume of traffic using the road. Maintenance costs are expected to be a constant \$200 000 per year of operation. The NPV of the road for various values of life and discount rate will be calculated.

Let the project life be n years and the discount rate $i\%$ p.a. The NPV of the road can be determined using Equations (5.7), (5.14) and (5.15):

$$\text{NPV(\$M)} = \frac{1.2}{(1+i)} + \frac{1.2\,(1.02)}{(1+i)^2} + \ldots + \frac{1.2\,(1.02)^{n-1}}{(1+i)^n} - \frac{0.2\,\{1-(1+i)^{-n}\}}{i} - 10$$

$$\text{NPV(\$M)} = \frac{1.2}{(1+i)}\left\{\frac{1-(1.02)^n/(1+i)^n}{1-(1.02)/(1+i)}\right\} \frac{-0.2\,\{1-(1+i)^{-n}\}}{i} - 10 \qquad (5.25)$$

Values of NPV for various discount rates and project lives are shown in Figure 5.17. Clearly for any given value of project life, the NPV decreases with increasing discount rate. For any given discount rate the NPV increases with increasing project life.

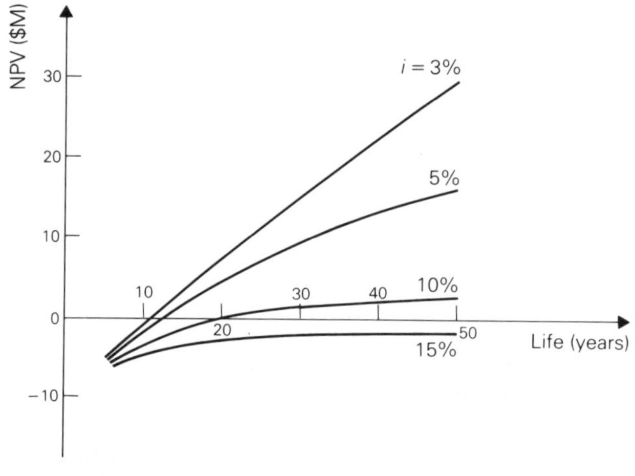

Figure 5.17 The effect of discount rate and life on the NPV of a road project.

It should also be noted that the economic desirability of the project depends strongly on the discount rate. For example, if i = 5%, NPV is positive provided the project life exceeds 12 years. On the other hand, for a value of discount rate of 15% the NPV is negative for all values of project life. For high values of discount rate, the NPV is relatively insensitive to changes in project life. For example, with a discount rate of 10% the NPV is $2.00M if the project life is 35 years and $2.67M if the project life is 50 years.

There has been considerable controversy associated with the selection of the discount rate (and project life), particularly for public sector projects. For example, consider the case of a public authority which is required to justify all projects on an economic basis and only undertake projects which have a positive NPV. Clearly the list of viable projects will be much greater if the authority uses a low value of discount rate than if a high value is used. This fact has lead to those who support large public works expenditure to argue for a low value of discount rate whereas those who wish to reduce public works expenditure often argue for a high value. Before discussing this matter further, we shall consider briefly the choice of discount rate for evaluating private sector projects.

The discount rate in the private sector

The correct concept for the discount rate is the **opportunity cost of capital**. The opportunity cost of capital for a particular company or individual at a particular point in time is the highest rate of return which it could obtain by having extra capital available. It obviously depends on the investment opportunities available at that time.

Most corporations raise their capital by a combination of borrowing from financial institutions (called **debt capital**) and reinvesting some of the company profits or retained earnings (called **equity capital**). In this case the opportunity cost of capital is a weighted average of the interest rate paid on debt capital and the rate of return which the shareholders could obtain by investing their capital to the best advantage elsewhere. This usually results in a higher rate than the interest paid on debt capital.

The discount rate in the public sector

As noted above the choice of a discount rate in the public sector is often controversial. The three most common theoretical bases for the choice of a discount rate are:

- the average yield on long-term government bonds
- the social opportunity cost
- the social time preference rate

As there is no agreed method we shall consider the arguments for and against each one in turn.

THE AVERAGE YIELD ON LONG-TERM GOVERNMENT BONDS

It has been argued that the interest rate paid on long-term government bonds is the rate at which the government borrows money from the public. It is therefore appropriate that it should be used as the discount rate for evaluating public sector investments. In Australia this rate is typically 3–4% p.a. above the inflation rate. However, if we examine the federal budget as a whole it is clear that most government revenue is raised by taxes, not by the sale of government bonds. In fact, bonds are used by the government to regulate the money supply rather than to raise revenue. The preceding argument is therefore questionable.

Another argument which supports the use of the long-term government bond rate is that this represents the risk-free interest rate in the private sector. Most investments carry some element of risk, due to uncertainties in the future economic climate. Even banks and building societies have been known to default on interest payments during major depressions. The greater the risk, the higher the interest rate expected by investors. The interest on government bonds is guaranteed by the government and so carries no risk. Any private investment should return at least this interest rate, otherwise investors would withdraw their money and put it into government bonds. It is argued that it is therefore appropriate to use it for public sector investments.

THE SOCIAL OPPORTUNITY COST

It is argued that funds invested by the government in major capital works have basically been displaced from the private sector and should, therefore, earn a rate of return at least as high as that prevailing in the private sector (based on the 'opportunity cost of capital' concept). However, not all revenue raised by government taxation would have been invested in the private sector. In fact a high percentage of government revenue would have been consumption expenditure and would have produced no investment return. This fact must be taken into account when assessing the social opportunity cost. A further difficulty is in determining the appropriate marginal rate of return in the private sector. A typical rate for the social opportunity cost is 10–15% above inflation.

THE SOCIAL TIME PREFERENCE RATE

It is argued that by undertaking investments in the public sector, society is effectively forgoing consumption in the current time period in order to achieve increased consumption in the future. The choice of a discount rate is equivalent to placing a weight on benefits and costs incurred in future years when compared with the present. Some economists consider that society is free to choose whatever stream of future consumption that it likes, and accordingly it may choose any set of relative weights on benefits and costs in future years. The discount rate corresponding to this chosen set of relative weights is the social time preference rate, which will in general differ from the interest rate prevailing in the private sector.

To illustrate the concept of social time preference rate, suppose that society (or its elected representatives) decides that benefits and costs received in one year's time should have a weight of W_1 relative to benefits and costs received in the present. Similarly, benefits and costs received in two years' time have a weight of W_2. In general, W_t is the weight placed on benefits and costs received at the end of year t relative to benefits and costs received in the present. (Clearly $W_0 = 1$.)

Then net community benefits (NCB) can be defined as:

$$\text{NCB} = \sum_{t=0}^{n} W_t\,(B_t - C_t) \tag{5.26}$$

Based on the preference of most individuals to consume now rather than later it is reasonable to assume that

$$W_0 > W_1 > W_2 > \ldots, \text{etc.} \tag{5.27}$$

A somewhat stronger assumption is that the weight placed on the benefits in any particular year is a constant fraction of the weight on the benefits in the preceding year, i.e.

$$W_1 = kW_0 = k$$

$$W_2 = kW_1 = k^2 \tag{5.28}$$

. . ., etc.

where k is a chosen parameter between 0 and 1.

Suppose we define k as follows:

$$k = \frac{1}{(1+r)} \tag{5.29}$$

where r is the **social time preference** (STP) rate.

Combining Equations (5.26), (5.28), and (5.29) we obtain

$$\text{NCB} = \sum_{t=0}^{n} \left\{ \frac{B_t}{(1+r)^t} - \frac{C_t}{(1+r)^t} \right\} \tag{5.30}$$

The similarity with Equation (5.18) should be noted. It should be clear that choosing the STP rate, r, is equivalent to choosing a set of weights on future benefits and costs for society as a whole.

The STP rate may be any rate chosen by society (even 0%). Proponents of the STP rate usually argue for a low rate (in the order of 3–5%) on the grounds that government investments provide basic community infrastructure and therefore benefits over a long time scale should be considered.

Ultimately the choice of the discount rate for public sector projects tends to be a political decision. Often sensitivity analyses are carried out in which the NPV for all projects is calculated using a low, medium, and high value of discount rate. For example, the NPV of all projects is determined using discount rates of 5%, 10%, and 15% p.a. and the results compared.

Project life

Many engineers consider the life of an engineering project to be equal to the physical life of the constructed facilities. The physical life of a system ends when it can no longer perform its intended function. For example, a building is at the end of its physical life when it is no longer fit for habitation. A road pavement is at the end of its physical life when it is suffering total breakup and can no longer be patched.

However, two other concepts of project life are relevant in economic evaluation. These are:

- the economic life
- the relevant period of analysis

The **economic life** of an engineering system ends when the incremental benefits of keeping the system operating for one more year are less than the incremental costs of maintenance and repair for one more year. As most engineering systems can be kept operating almost indefinitely with suitable replacement of parts, the economic life is usually less than the physical life.

The **relevant period of analysis** is the maximum period beyond which the system performance has virtually no effect on its present value of benefits and costs. For high values of discount rate the NPV of a project does not significantly increase beyond a certain time. This time is the **relevant period of analysis**. For example, if a discount rate of 15% is being used it is pointless arguing whether the life of a building will be 30 years or 100 years as it will make no difference whatsoever in its economic evaluation. For lower values of discount rate more care must be taken in choosing the project life.

In general, the project life is the smallest of the physical life, the economic life and the relevant period of analysis.

Some typical lives used for civil engineering projects are given in Table 5.11.

Table 5.11 Lives of some engineering systems.

Vehicles	5–10 years
Roads	20–30 years
Bridges, buildings	30–50 years
Dams	50 years
Tunnels, cuttings, embankments	100 years

5.10 Case study

A new freeway is being planned to run between two cities. The present road has a length of 50 km and an annual maintenance cost of \$3000/km. It is currently in need of a major overhaul which will cost \$5M.

The proposed new freeway will cost \$60M to construct. It will have a length of 40 km and an annual maintenance cost of \$6000/km.

Other relevant data for the two roads are given in Table 5.12.

The state road authority has decided that a discount rate of 6% and a project life of 30 years are appropriate for this project. The problem is to determine if the proposed freeway is justified on economic grounds.

In reality other factors would need to be considered. These include the economic evaluation of competing projects, political considerations, and environmental and social issues. From the viewpoint of economic evaluation alone, this decision is basically a go or no-go decision. Either the new freeway will be built or the existing road overhauled. Therefore, the decision will be insensitive to the economic criterion used.

Table 5.12 Data for existing road and proposed freeway.

Variable	Existing	Proposed
Traffic volume (vehicles per day)		
– Commercial	500	600
– Non-commercial	4000	5000
Average speed (km/h)	80	100
Vehicle operating costs (cents/km)		
– Commercial	40	30
– Non-commercial	10	12
Value of travel time ($/h)		
– Commercial	16	16
– Non-commercial	6	6
Average number of accidents/year	30	6
Average cost per accident ($)	10 000	5000

The state road authority would probably make use of either NPV or B/C ratio, so both will be used for comparison. It is important to realize that the evaluation will be of the proposed freeway *in comparison* with the existing road. It is pointless trying to determine whether the existing road is economically viable or not, as the possibiity of providing no road is not at issue. Following the principles outlined in Section 5.7, all benefits and costs of the proposed freeway will be *incremental* to the existing road. The vehicle operating cost figures in Table 5.12 represent the marginal cost of fuel, oil, maintenance, and tyres. Fixed costs such as registration and insurance are excluded, as these will be paid in any case. The value of travel time used represents willingness-to-pay for savings in time. For commercial operators this should at least equal the drivers' wage rate plus overheads. For non-commercial drivers it is their willingness-to-pay for additional leisure time.

Benefits
The benefits of the proposed highway are: savings in travel time, savings in operating costs, and reduction in accident costs. The average travel time on the existing road is:

$$\frac{50\,\text{km}}{80\,\text{km/h}} = 0.625\,\text{h}$$

with the freeway this becomes:

$$\frac{40\,\text{km}}{100\,\text{km/h}} = 0.4\,\text{h}$$

The total (i.e. travel time plus operating cost) per vehicle trip for commercial vehicles on the existing road is:
0.625 h × \$16/h + 0.40 \$/km × 50 km = \$30/vehicle trip
On the freeway it is:
0.4 h × \$16/h + 0.30 \$/km × 40 km = \$18.40/vehicle trip

In response to this reduced cost, it is anticipated that the volume of commercial vehicles will increase from 500 vehicles/day to 600 vehicles/day. The benefits to commercial users will be given by the increase in consumers' surplus as shown in Figure 5.18.

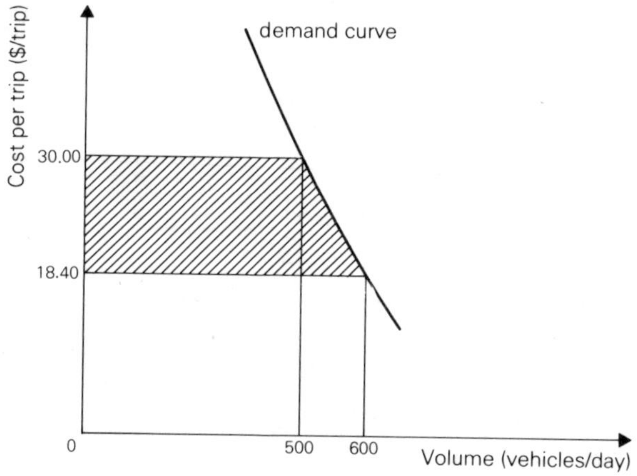

Figure 5.18 Increase in consumers' surplus for commercial vehicles.

Assuming the demand curve is approximately a straight line between 500 and 600 vehicles/day, the daily benefit is given by Equation (5.22), as follows:
Daily benefit to commercial users:
½(500 + 600) × (30 − 18.4) = \$6380/day
Therefore the annual benefit to commercial users is:
6380 × 365 = \$2.33M/year
For non-commercial vehicles, the total cost/trip on the existing road is:
0.625 h × \$6/h + 0.10 \$/km × 50 km = \$8.75/vehicle
On the freeway this becomes:
0.4 h × \$6/hour + 0.12 \$/km × 40 km = \$7.20/vehicle
The daily benefits to non-commercial users can similarly be determined to be
½(4000 + 5000) × (8.75 − 7.20) = \$6980/day
The annual benefit to non-commercial users is thus:
6980 × 365 = \$2.55M/year
The savings in accident costs/year are:
30 × 10 000 − 6 × 5000 = \$270 000/year

The total annual benefits of the freeway relative to the existing road are therefore

$$\begin{array}{rl} \$2.33\text{M} & \text{(commercial users)} \\ +2.55\text{M} & \text{(non-commercial users)} \\ +0.27\text{M} & \text{(accident reduction)} \\ \hline = \$5.15\text{M} & \end{array}$$

Costs
Incremental capital cost = \$60M − \$5M = \$55M
Incremental annual maintenance cost

$$= 40\,\text{km} \times \$6000/\text{km} - 50\,\text{km} \times \$3000/\text{km}$$
$$= \$90\,000/\text{year}$$

Using Equations (5.7) and (5.8), we thus have
PVB = 5.15 (*P*/*A*,6%,30)

$$(P/A, 6\%, 30) = \left\{\frac{1-(1.06)^{-30}}{0.06}\right\} = 13.76$$

PVB = 5.15M × 13.76 = \$70.86M
PVC = 55M + 90 × 10^3 × 13.76 = \$56.24M
NPV = (70.86 − 56.24) = \$14.62M

$$\text{B/C} = \frac{70.86}{56.24} = 1.26$$

The new freeway is therefore justified on economic grounds, according to both the NPV and B/C criteria. As mentioned earlier other factors would need to be considered before deciding whether the project should proceed. These include the economic evaluation of other competing projects, political considerations and environmental and social issues. The question of environmental and social evaluation is addressed in Chapter 6.

Appendix: Price determination in perfectly competitive markets

The **perfectly competitive market** is a theoretical model used by economists to analyse market behaviour in much the same way that engineers use free-body diagrams to analyse the forces on a group of rigid bodies.

It is hypothesized that there is a market of buyers and sellers which determines the market price for any particular good or service (called a commodity).

The assumptions inherent in the perfectly competitive market are:

- there are a large number of buyers and sellers of the commodity
- any particular commodity is identical for all sellers

- all buyers and sellers have perfect information about prices and production costs
- there is free entry and exit of firms into the market

The market supply curve

For any particular commodity there is a **market supply curve** which shows the total amount of the commodity which will be produced per unit time at any particular price. For example, Figure 5.19 shows the market supply curve for timber in a particular economy. (Note that it is traditional to plot price on the vertical axis.) As price increases we would expect a greater production of timber as more producers come into the market and existing producers increase their output by working overtime, investing in new machinery, and so on.

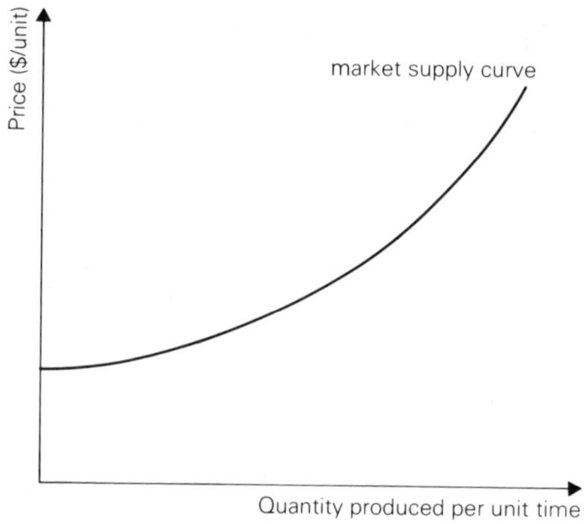

Figure 5.19 Market supply curve for timber.

Other factors apart from the price of the commodity can affect the quantity supplied. In terms of Figure 5.19 these factors will cause a shift in the position of the market supply curve. These factors are:

- **the price of inputs** (for example, labour or materials) to the production process
- **the price of competing outputs.** For example, an oil refinery can be used to produce many products including petrol and kerosene. If the price of petrol increases relative to the price of kerosene, the market supply curve for kerosene would tend to shift to the left
- **the state of technology.** Technological advances can reduce production costs and cause the market supply curve to shift to the right
- **the time period.** The market supply curve for a short time period tends to be near vertical, as there is little time to respond to changes in price. In

the longer term, the curve tends to be flatter as there is more time for producers to respond to changes in price
- **expectations.** The quantity supplied at any particular price will depend on whether producers expect prices to increase, decrease or remain the same in the next time period

The market demand curve

For any particular commodity there is also a **market demand curve**. This shows the quantity per unit time of the commodity which would be purchased at any particular price. Figure 5.20 shows the market demand curve for timber in a particular economy. As the price of timber increases we would expect the demand to fall as consumers use substitute materials such as plastic, steel, and aluminium.

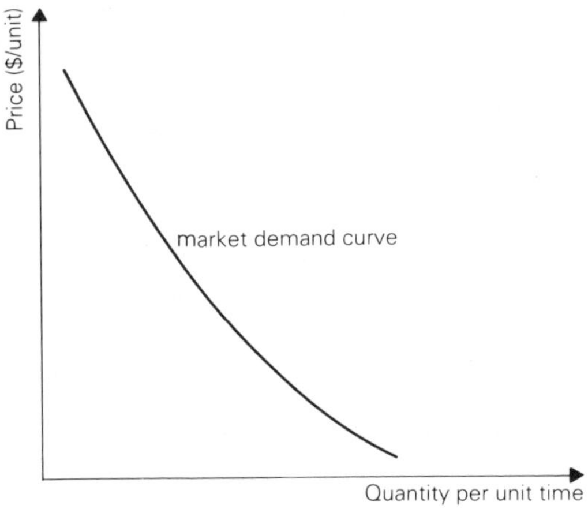

Figure 5.20 Market demand curve for timber.

Apart from price, a number of other factors affect the position of the market demand curve. These include the following:

- **the prices of other commodities.** Certain commodities are substitutes for one another. Consider, for example, gas and electricity. If the price of gas increases, the market demand curve for electricity tends to shift to the right.
  Other commodities such as concrete and steel reinforcing rods are complementary goods. If the price of concrete increases there is a shift to the left of the demand curve for steel reinforcing rods.
- **income and wealth of consumers.** If there is a general rise in income levels there is an increase in demand for all normal goods. This is reflected by a shift to the right of the demand curve.
- **number of consumers.** An increase in the number of consumers causes the market demand curve to shift to the right.

- **tastes of consumers.** Changes in taste and fashion may cause a shift in the demand curve for some commodities, e.g. clothing.
- **the time period and expectations.**

Market equilibrium

The market price for any particular commodity in a perfectly competitive market is determined by the interaction of demand and supply. At equilibrium the market price corresponds to the point of intersection of the market supply and market demand curves. This point is illustrated in Figure 5.21, where the equilibrium price is given by P^* and the equilibrium quantity by Q^*.

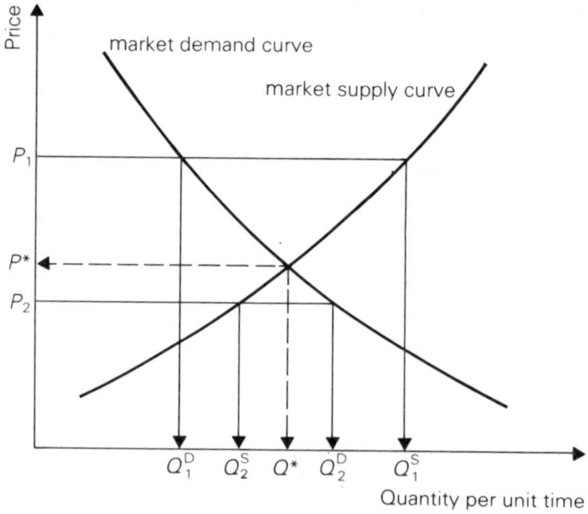

Figure 5.21 Demand–supply equilibrium.

Suppose the price is set at a value greater than P^*, for example P_1. Then the quantity supplied, Q_1^S exceeds the quantity demanded, Q_1^D. The product therefore stockpiles and there is pressure on producers to reduce the price in order to sell their product.

If the price is below P^* (e.g. P_2), the quantity demanded, Q_2^D exceeds the quantity supplied, Q_2^S, and there are shortages of the commodity. In response to these shortages, consumers will tend to bid up the price of this scarce commodity. As the price increases, producers will tend to produce more and new producers will enter the market.

At a price of P^* the quantity demanded just equals the quantity supplied and an equilibrium state is established.

For example, consider the demand/supply equilibrium of traffic on a section of urban arterial road. In this case, each consumer may decide either to make one or more trips on the road or not to use it at all. The *price* of a trip is the total cost of petrol, oil, and tyre wear of making one traverse of the section of road. To this the value of travel time of the driver and passengers should be added. The *quantity* is that number of vehicles per hour travelling on the road.

Figure 5.22 shows supply and demand curves for the road. The supply curve indicates the total cost per trip as a function of the volume of traffic on the road in vehicles per hour. Typically, the cost per trip increases rapidly as the volume of traffic approaches the capacity of the road, *C*. This is due to the effects of traffic congestion and delays.

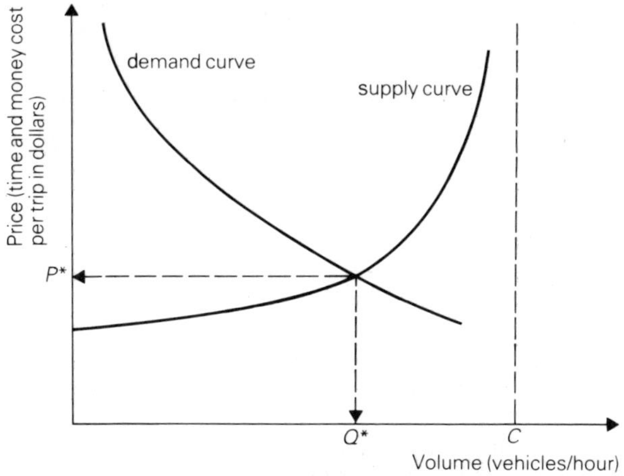

Figure 5.22 Demand–supply equilibrium for an urban arterial road.

The demand curve in Figure 5.22 indicates the number of drivers per hour wanting to use the road as a function of the total cost per trip. As the cost per trip increases, fewer drivers wish to use the road because of available substitutes such as other routes of travel or other times of the day.

If the traffic conditions on the road remain steady for a sufficient period of time, equilibrium is reached with a volume of Q^* vehicles per hour on the road. The corresponding cost per trip is P^* dollars.

Shifts in demand and supply curves

As mentioned earlier, the demand curve for a particular commodity may shift with time due to changes in consumer preferences, the number of consumers, average income levels or the prices of related goods. A shift in demand leads to a new equilibrium quantity and price for the commodity.

For example, suppose the demand for water in a city increases due to an increase in population. This causes the demand curve for water to shift to the right as shown in Figure 5.23. If the water supply authority charges the market price for water, one would expect the volume supplied to increase from Q_1 to Q_2 and the price of water to increase from P_1 to P_2 as less economical sources of water are used.

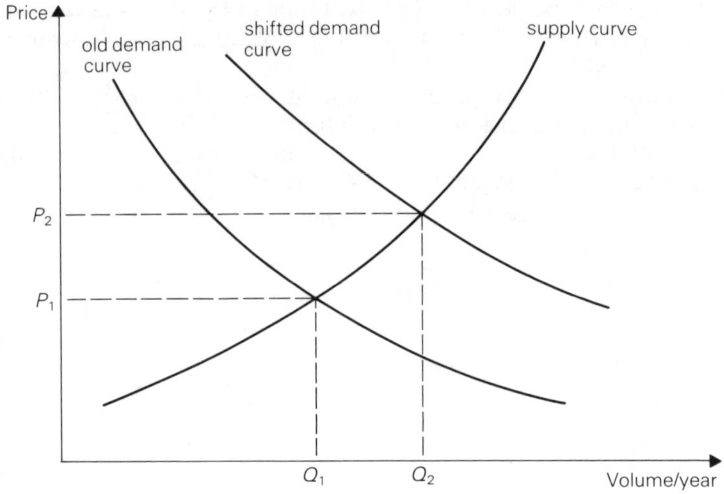

Figure 5.23 A shift in the demand curve for water.

In a similar fashion, the construction of a new engineering system causes the supply curve to shift to the right and a new equilibrium to be established.

For example, in response to increasing demand, a water supply authority builds a new dam to supply water to a city. This causes the supply curve to shift to the right as more water can now be supplied at any particular price. The situation is shown in Figure 5.24. In equilibrium the volume of water supplied increases from Q_1 to Q_2 and the price falls from P_1 to P_2. In practice, this fall in price may take place over several years by allowing the

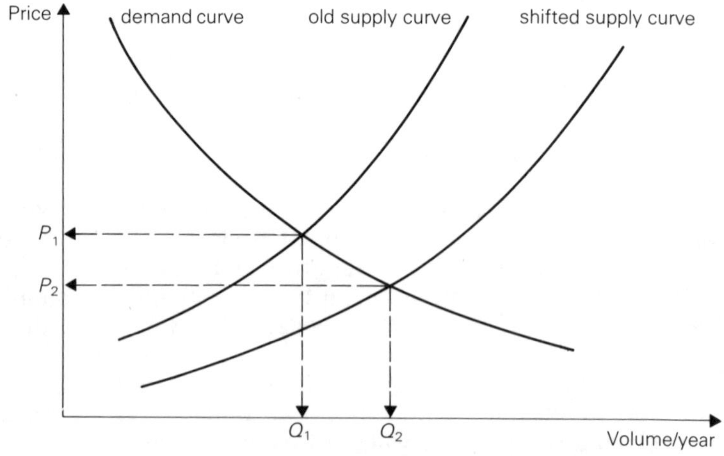

Figure 5.24 A shift in the supply curve for water.

rate of increase of the nominal price of water to fall behind the rate of inflation.

Price elasticity of demand

In planning future engineering facilities such as transportation and water resource systems, the analyst is often concerned with the responsiveness of consumers to changes in price.

It is possible to characterize the market demand for a product at any point in time by its price elasticity of demand. This is defined as the percentage change in quantity demanded, divided by the percentage change in price, i.e. **elasticity**,

$$\varepsilon = \frac{\Delta Q/Q}{\Delta P/P} = \frac{\Delta Q}{\Delta P}\left(\frac{P}{Q}\right) \tag{5.31}$$

For example, the demand for water for irrigation from a particular river is $1.2 \times 10^9\,\text{m}^3$ per year when the price of water is $0.10/m³. If the price increases to $0.15/m³ and the price elasticity of demand is −0.3, what will be the new demand? Using Equation (5.31),

$$\varepsilon = -0.3 = \frac{\Delta Q}{0.05} \times \frac{0.10}{1.2 \times 10^9}$$

therefore

$$\Delta Q = \frac{-0.3 \times 0.05 \times 1.2 \times 10^9}{0.10} = -0.18 \times 10^9\,\text{m}^3/\text{year}$$

$$\text{New demand} = 1.2 \times 10^9 - 0.18 \times 10^9$$

$$= 1.02 \times 10^9\,\text{m}^3/\text{year}$$

Strictly speaking, the price elasticity defined by Equation (5.31) is the **arc elasticity**. In the limit as ΔP approaches zero, ε becomes the **point elasticity**, ε_P. This is defined as

$$\varepsilon_\text{P} = \frac{\text{d}Q}{\text{d}P}\left(\frac{P}{Q}\right) \tag{5.32}$$

For a downward-sloping demand curve the elasticity is negative. It is common to ignore the minus sign when quoting values of the price elasticity of demand.

If the price elasticity of demand is less than minus one the demand is said to be **elastic**, otherwise the demand is called **inelastic**.

Some typical estimated values of price elasticities of demand of interest to engineers are given in Table 5.13.

Table 5.13 Some typical values of price elasticity of demand.

Commodity	Price elasticity	Reference
Water, for in-house use	−0.23	Howe and Linaweaver (1967)
Water for outdoor use,		
western USA	−0.70	
eastern USA	−1.57	
Intercity passenger travel (north-east USA),		
Business:		
Rail	−0.35	Systems Analysis and Research Corporation (1963)
Bus	−0.74	
Air	−0.89	
Car	−0.36	
Personal:		
Rail	−3.00	
Bus	−0.69	
Air	−0.91	
Car	−0.93	

References

Baumol, W. J. 1977. *Economic theory and operations analysis*, 4th edn. Englewood Cliffs, NJ: Prentice-Hall.

DeGarmo, E. P., W. G. Sullivan & J. R. Canada 1984. *Engineering economy*, 7th edn. New York: Macmillan.

deNeufville, R. & J. H. Stafford 1971. *Systems analysis for engineers and managers*. New York: McGraw-Hill.

Howe, C. W. & F. P. Linaweaver 1967. The impact of price on residential water demand and its relation to system design and price structure. *Water Resources Research* **3,** 13–32.

Jackson, J. & C. R. McConnell 1980. *Economics (Australian Edition)*. Sydney: McGraw-Hill.

Lipsey, R. G., P. C. Langley & D. M. Mahoney 1981. *Positive economics for Australian students*. London: Weidenfeld & Nicolson.

Meredith, D. D., K. W. Wong, R. W. Woodhead & R. H. Wortman 1985. *Design and planning of engineering systems*, 2nd edn. Englewood Cliffs, NJ: Prentice-Hall.

Mishan, E. J. 1982. *Cost–benefit analysis*, 3rd edn. London: George Allen & Unwin.

Ossenbruggen, P. J. 1984. *Systems analysis for civil engineers*. New York: Wiley.

Riggs, J. L. 1977. *Engineering economics*. New York: McGraw-Hill.

Sage, A. P. 1982. *Economic systems analysis. Microeconomics for systems engineering, engineering management and project selection*. Amsterdam: Elsevier.

Smith, A. A., E. Hinton & R. W. Lewis 1983. *Civil engineering systems analysis and design*. London: Wiley.

Steiner, H. M. 1980. *Public and private investments*. New York: Wiley.

Systems Analysis and Research Corporation 1963. *Demand for intercity passenger travel in the Washington–Boston corridor*. Report No. PB 166884, NTIS, Springfield VA, USA.

Problems

5.1 A company borrows $300 000 at an interest rate of 13.5% p.a. (compound). What constant annual payment must be made at the end of each year in order to pay the loan off in 15 years?

5.2 An engineer wishes to have $500 000 in a bank account when she retires in 30 years time. What constant amount should she invest at the start of each year if the interest rate on the investment is 11% p.a. compound?

5.3 An engineering contractor owes $36 000 on a loan which has an interest rate of 20% p.a. (compound). He is currently making payments of $9000 per year on the loan.

(a) How many years will it take to pay off the loan if he keeps up his current annual payments?
(b) What will be his final payment on the loan?

5.4 You have $10 000 to invest for 1 year. Which of the following alternatives will give the highest return:

(a) 12% p.a. compounded annually
(b) 11.75% p.a. compounded monthly
(c) 11.5% p.a. compounded daily

5.5 A man has to decide whether to install solar collectors on his domestic hot water service. The cost of the solar system is $600 and he estimates that it will reduce his annual electricity bill for water heating by 50%. The solar collectors have a life of 10 years. Currently he pays $160 p.a. for water heating. If the expected annual increase in electricity charges is 10% p.a. (compound), are the solar collectors a good investment under the following conditions:

(a) He has nothing in the bank and would have to borrow money at 18% p.a.
(b) He has $2000 in a savings investment account earning 13% interest p.a.

5.6 A manufacturing company has space for one additional automated machine in its factory. It can purchase either machine A which produces gadgets or machine B which produces widgets. Economic data for the two machines are presented in Table 5.14.

(a) Compute the net present value, equivalent annual worth, B/C ratio and internal rate of return for each machine using a discount rate of 7% per annum.
(b) Which machine should be chosen and why?

Table 5.14 Economic data for two machines.

	Machine A	Machine B
Initial cost ($)	310 000	370 000
Annual revenue generated ($/year)	70 000	120 000
Annual maintenance costs ($/year)	15 000	20 000
Life (years)	10	5
Scrap value ($) (at end of life)	20 000	0

5.7 A new town is being planned to service a proposed mining development in a remote area. Two alternative sources of water supply to the town are being considered. These are:

- Groundwater pumped from a nearby aquifer.
- The construction of a dam in a range of hills some distance from the town and a pipeline connecting to the town.

Relevant economic information for the two alternatives is listed in Table 5.15.

Table 5.15 Economic data for two water supply systems.

	Groundwater scheme	Dam and pipeline
Initial cost ($)	150 000	1 000 000
Annual operating cost ($)	100 000	40 000
Annual benefits ($)	200 000	240 000
Life (years)	10	30

It is anticipated that the mine will operate for 20 years, after which the town, mine, and water supply system will be abandoned.

(a) Using a discount rate of 8% p.a. compare the two water supply alternatives in terms of net present value.
(b) Calculate the internal rate of return of the dam and pipeline project.

5.8 A new bridge is to be built and it will be funded entirely out of toll revenue. The bridge will cost $120 million to build. Annual maintenance and toll collection costs are expected to be $1 million in the first year of operation and these will increase by 5% per year in a compound fashion. The volume of traffic which will use the bridge is estimated to be an average of 50 000 vehicles per day for the first year of operation. This will increase by 2% per year in a compound fashion.

If the bridge authority takes an initial loan of $120 million at 15%

compound interest per year, which must be repaid in 40 years, what toll per vehicle should be set for the bridge?

(Assume the toll remains constant for a 40-year period and that, each year, the authority pays any surplus funds towards the loan.)

5.9 A state water authority currently supplies 400 000 ML of water per year for irrigation purposes at a price of \$100/ML. The authority is considering constructing a dam which will increase the available supply by 180 000 ML per year.

(a) If the arc elasticity of demand for water is −0.4, what price should the water authority charge for water to ensure that the available supply is all used?
(b) Estimate the economic benefits of the increased supply.

6 Environmental and social considerations

In this chapter the environmental and social effects of engineering projects are considered. This is a complex field in which engineers often need to call on the expertise of other professionals. Engineers should, however, be aware of the issues involved and be familiar with techniques for carrying out environmental and social evaluation.

6.1 Introduction

Most large engineering projects have a number of impacts on the physical and social environment. Some of these are beneficial and others harmful. For example, the construction of a large dam could have a harmful effect on the fish and other organisms which live in the river downstream of the dam due to a change in the flow regime. The reservoir itself could flood an area of bushland which provides a habitat for many terrestrial animals and birds. However, the artificial impoundment above the dam could provide a suitable habitat for other fish and waterfowl as well as providing increased recreational opportunities for picnics and sailing.

Engineers have a special role to play in the wise use, conservation and management of the earth's resources. The nineteenth-century view of engineering as '. . . the art of directing the great sources of power in nature for the use and convenience of man . . .' (see Section 1.2) has been tempered by a realization that we live in a complex world of finite resources. It is only by learning to live in harmony with nature and each other that we can ensure the long-term survival of the human race. The engineering profession plays a major role in environmental protection and conservation. Some engineers coordinate environmental impact studies, while others develop new technology to reduce waste emissions to the air, water, or land. Still other engineers are involved in the management of our water, soil, and energy resources.

Social effects have recently received increased attention by the engineering profession as it is recognized that many engineering projects have a range of effects on various groups in the community. A new urban freeway, for example, may provide reduced operating costs and travel times to the users, but zero benefit to those who are too old or too poor to drive a car. Furthermore, it might provide large disbenefits to inner suburban dwellers who live near the freeway and must suffer increased noise, air pollution, and intrusion. Thus the distribution of benefits and costs in the community can be an important factor in deciding the acceptability of an engineering project.

6.2 Environmental quality

Various definitions of the environment have been proposed. One reasonable definition is as follows (IEA 1987):

> 'The total stock of social, physical and biological resources available to man and to other species, and the ecosystems of which they are part.'

Environmental quality considerations are important in the evaluation of any large engineering project. In the private sector, environmental considerations tend to enter the planning or design process as **constraints**. Many governments require potential developers to provide an **environmental impact statement** for large-scale developments. The intention of preparing this statement is to ensure that the environmental (and social) effects of development are kept within acceptable bounds. Engineers often play a coordinating role in the preparation of environmental impact statements. A private developer must ensure that a proposed project does not exceed acceptable levels for emissions to air, water, and soil, ecological effects, social disruption, and so on. These constraints are imposed and monitored by government bodies.

In the public sector, environmental quality can be viewed as a set of **objectives** rather than a set of **constraints**. In many cases, government projects aim to achieve high levels of environmental quality often at the expense of economic benefits. An example of this philosophy is embraced in the U.S. 'Principles and Standards for the Planning of Water and Related Land Resources' (USWRC 1973). The Principles and Standards were a directive from the President of the USA to federal government agencies and those seeking federal funding for water resources developments. The Principles and Standards required the responsible agencies to plan such developments so as to maximize two sets of objectives of equal importance, namely:

- National economic development (NED)
- Environmental quality (EQ)

In addition 'regional development' and 'social well-being' were mentioned for possible consideration in the assessment procedure, although it was not necessary to develop plans which maximized these objectives.

The Principles and Standards have now been replaced by the 'Economic and Environmental Principles and Guidelines for Water and Related Land Resources Implementation Studies' (USWRC 1983). The 1983 Principles and Guidelines emphasize the NED objective, consistent with protecting the environment. This is equivalent to maximizing economic efficiency subject to constraints on acceptable environmental quality.

Even if not bound by such strict guidelines, engineers need to be sensitive to environmental concerns and the likely impacts which their systems may have on the physical and social environment. They should be prepared to take action to ameliorate these impacts where appropriate.

Clearly there are many parameters of environmental quality which could be affected by any large-scale engineering development.

One list of environmental parameters relevant to water resource projects is shown in Figure 6.1. This forms part of the Battelle environmental evaluation system (Dee *et al*. 1973). In this scheme, four broad categories of environmental quality are recognized. These are **ecology**, **environmental pollution**, **aesthetics**, and **human interest**. Each category is divided into components, e.g. environmental pollution is divided into water pollution, air pollution, land pollution, and noise pollution. Each component is, in turn, divided into a number of parameters.

We will consider each of the four categories in turn.

The ecology category includes all plants and animals in the natural environment. Environmental effects can occur at the population, community, or ecosystem level.

A **population** is a group of individuals of the one species occupying a particular area. As shown in Figure 6.1, consideration needs to be given to the effects of a project on both terrestrial (land-based) and aquatic (water-based) populations.

The next level of ecological organization is the **community**, which is a collection of populations living in a prescribed area. It is important to recognize that animals and plants are not randomly scattered, but are grouped into communities of interacting populations. The interactions which occur within a community include predation, parasitism, competition, and mutualism (in which both species benefit). The community concept enables us to recognize that an effect on any one population will have an impact on other populations in the community. For example the clearing of trees from a catchment may impact on the populations of animals, birds, and other plants forming part of that community.

An indication of the health of a community is its **species diversity**. Mature communities tend to be more diverse and more stable. Low diversity is usually the sign of a community which is under stress, either due to natural variations (e.g. in the intertidal zone) or high pollution levels.

A **food web** for a community is a diagram which illustrates which species eat which other species. In a similar way to species diversity it is an indicator of the health of a community, with simpler food webs being associated with stressed communities.

An **ecosystem** consists of all the organisms and all of the non-living matter and energy which interact in a given area. The systems concepts outlined in Chapter 3 apply equally well to an ecosystem as to an engineering system. A lake, a river, and a forest are examples of ecosystems. The ecosystem is often the most appropriate level at which to assess the effects of manmade changes on the environment.

The ecology category in the Battelle System requires evaluation at the population, community, and ecosystem level.

The environmental pollution category in Figure 6.1 consists of water, air,

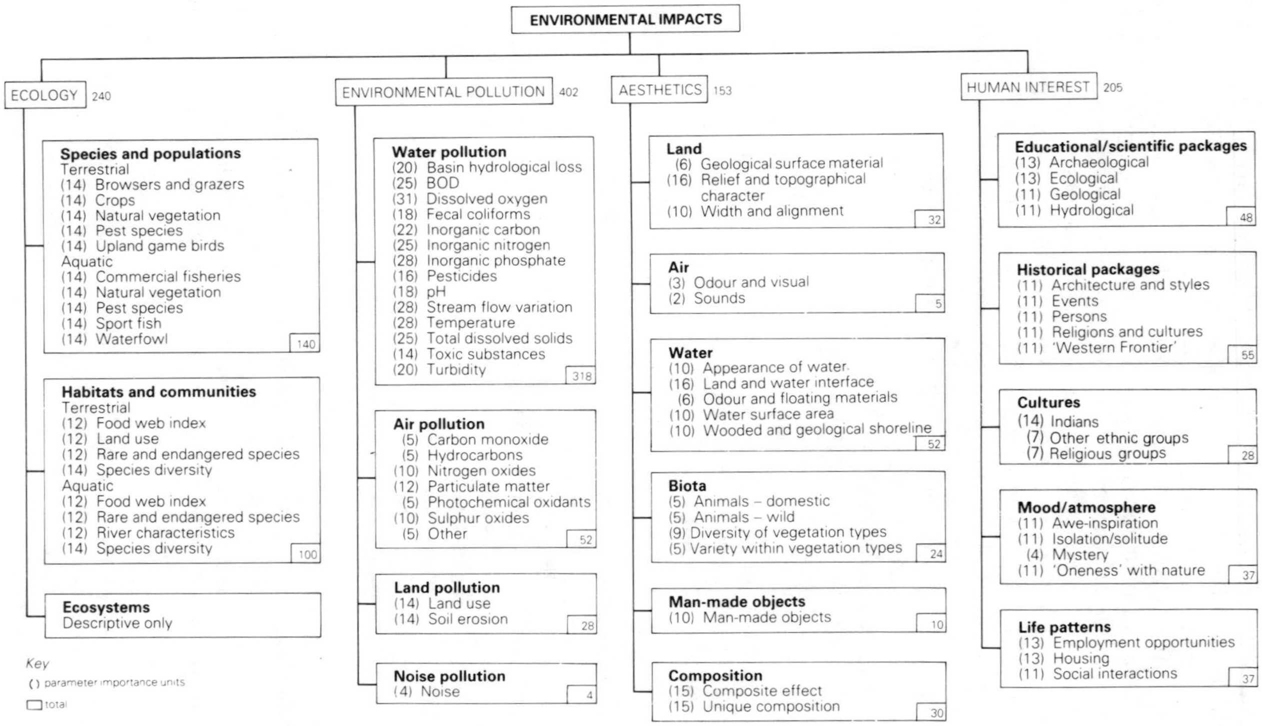

Figure 6.1 Environmental evaluation system (Dee *et al*. 1973).

land, and noise pollution. Some of the parameters listed are of direct concern for human health. These include faecal coliforms and toxic substances in water as well as carbon monoxide, nitrogen oxides, and sulphur oxides in air. Others such as noise are of concern because they may annoy members of the community. Still others such as dissolved oxygen, inorganic nitrogen and phosphate, and temperature in water are of concern because of the possible adverse impact on the biological community or ecosystem.

Even though the primary emphasis of engineering design is on functionality, aesthetics are also an important consideration. The Battelle evaluation system includes various parameters to assess the aesthetic impact of water resource developments (see Figure 6.1). These parameters are very difficult to quantify, and ultimately judgement is required in their assessment.

The human interest category includes parameters related to scientific discovery, historical, and cultural factors. These overlap with the social objectives discussed in the following section.

Clearly engineers are not expert in all of the areas relevant in an environmental investigation and must therefore call on specialist advice from ecologists, chemists, landscape architects, and others.

Every environmental parameter has its own scale(s) of measurement. Some of these are quantitative (e.g. the dissolved oxygen level in a river measured in milligrams per litre) others are qualitative (e.g. the aesthetic value of a dam). It is preferable, where possible, to use cardinal scales of measurement. The distinction between these and ordinal scales is discussed in the Appendix to this chapter.

6.3 Social well-being

The question of whether social effects should be considered as part of environmental quality or under a separate category is a matter of semantics. An environmental impact statement usually includes social impacts as well as effects on the physical and biological components of the environment. On the other hand, the Principles and Standards (USWRC 1973) considered social well-being as a separate category of objectives.

The social objective has been defined as follows:

> 'To enhance social well-being by the equitable distribution of real income, employment and population, with special concern for the incidence of the consequences of a plan on affected persons or groups; by contributing to the security of life, health and property; by providing educational, cultural and recreational opportunities; and by contributing to national security.' (USWRC 1973.)

If a full economic, environmental and social evaluation of a project is carried out, it is most important to avoid **double-counting** of benefits or costs under more than one account. If, for example, an economic value is placed on lives saved through accident reduction and this is included in the *economic* evaluation of the project, it would not be appropriate also to highlight the number of lives saved in the *social* evaluation. Other examples of double-counting are discussed later in this section.

The following types of social effects may need to be considered for a particular project:

- distribution of income
- population distribution
- employment
- life, health, and safety
- educational, cultural, and recreational opportunities
- national security and emergency preparedness
- displacement and relocation
- neighbourhood disruption and intrusion

As with environmental effects, each social parameter can have its own ordinal or cardinal scale of measurement. The above effects are considered in turn.

DISTRIBUTION OF INCOME

In theory it is possible for the government to redistribute the benefits and costs of any public sector project among the various groups in society. This being the case, it can be argued that the role of the engineer or planner is to devise projects which maximize the total net benefits to the community and let the government decide how these benefits (and costs) will be distributed. However, this redistribution of benefits and costs is often not practicable because of the difficulties in identifying all affected groups and because of the administrative cost of carrying out the transfers. Redistribution of the benefits of engineering projects rarely occurs in practice.

It is therefore important that the distributional consequences of any major project be considered explicitly. Of particular relevance is the effect of the project on the income levels of certain target groups. These target groups may be distinguished on the basis of income (e.g. those living below the poverty level), race, sex, or geographic region.

It has been suggested by Weisbrod (1972) that the distributional question can be approached by placing explicit weights on the benefits and costs received by each group in the community. Traditional benefit–cost analysis gives the same weight to a dollar of benefits received by a pauper and by a millionaire. It is conceptually possible to give a higher weight to benefits or costs incurred by low-income earners and thus develop a single index of social welfare, using the following formula:

$$\mathrm{SW} = \sum_{i=1}^{N} w_i \,(B_i - C_i) \tag{6.1}$$

where SW is the index of social welfare, B_i is the present value of benefits received by group i, C_i is the present value of costs incurred by group i, w_i is the 'social' weight of benefits or costs to group i, and N is total number of community groups.

The difficulties of deciding on a set of weights and implementing this approach are apparent.

POPULATION DISTRIBUTION

It has been argued that the redistribution of population throughout a region or the nation may be a major objective of government policy. This is particularly true in sparsely populated nations such as Australia. Major engineering projects in remote areas may contribute positively to the decentralization of population. The benefits of decentralization are related to enhanced national security (or defence) and the benefits of rural versus urban living.

High concentrations of population and industry are particularly vulnerable to attack by conventional or nuclear weapons. On the other hand, the settlement of remote areas leads to additional infrastructure costs for electricity supply, roads, railways, water supply, ports, and airfields.

Projects such as major irrigation developments which contribute to maintaining the size of the rural community may be supported on the grounds of the 'healthier' lifestyle of country living. Continued growth of large cities at the expense of the surrounding country areas may also involve considerable increases in infrastructure costs in the city, as well as increases in the crime rate and other social problems.

The negative aspects of population redistribution on individual families also need to be considered. These include the weakening of friendships and family ties and the stress associated with moving, setting up a new home, and adjusting to a changed lifestyle.

EMPLOYMENT

A large engineering project may involve considerable employment in the construction phase and also in the operations phase. Some people consider this an achievement in itself. In an economy with little or no unemployment, workers employed on a specific project will be displaced from productive activities elsewhere in the economy. Therefore, the opportunity cost of their labour will be assessed as a *cost* to the project in the economic evaluation. In such a case, the number of people employed on the project is not a social benefit.

In an economy with high unemployment, some of the workers employed on the project may have been previously unemployed. As outlined in Section 5.7, previously unemployed labour has an opportunity cost to the project considerably less than the wage rate. If this effect is correctly taken into account in the economic evaluation of all projects, there is no need to further describe the employment implications of the projects, as this would be a double-counting of the same benefit.

The foregoing applies to public sector projects only. In the economic evaluation of private sector developments all labour is costed at the actual wage rate paid. Thus at times where there is less than full employment in the economy, the true net benefits of the project to society will be understated.

In such cases it *is* valid to include the additional employment opportunities created by the project as part of the social evaluation.

The distributional consequences of employment opportunities may also be important; for example, in the creation of employment opportunities in depressed regions of the country or among disadvantaged groups such as women, the handicapped, ethnic, or low-income groups.

LIFE, HEALTH, AND SAFETY

Many engineering works contribute to saving lives, improving health, or increasing safety in the community. Typical examples are flood mitigation works, water treatment plants, sewage treatment, improved road alignments, grade separation, and coastal protection works.

Attempts have been made to include the benefits of lives saved into the economic evaluation of public sector projects by placing an economic value on human life.

Various methods have been suggested for estimating the economic benefits of saving human life. These include:

- The present value of the person's expected future earnings.
- A value imputed from political decisions which involve public expenditure aimed at reducing the number of deaths. For example, the government spends \$50M in equivalent annual worth (see Section 5.4) to upgrade railway crossings which will save an estimated 100 lives per year from fatal accidents. The implied value of saving one life is greater than or equal to \$50M/100 (i.e. \$0.5M).
- A dollar amount which people are prepared to accept as compensation in order to put up with the additional risk of death involved in the project. For example, a proposed new chemical plant will increase the probability of death per year to each person in an adjoining town by 1 in 10 000. The economic cost of this increased risk is the sum over all people in the town of the minimum level of compensation which each person is prepared to accept in order to tolerate the increased risk. The same basis can be used for projects which *reduce* the probability of death for part of the community.

Practical and theoretical difficulties exist with all of these economic approaches. An alternative approach is to estimate the number of lives saved per year and treat this as a social benefit.

Similar arguments apply to projects which increase or reduce the risk to public health or the chance of accidents occurring. Either a dollar value may be used in the economic evaluation or an estimated reduction in the risk may be included in the social evaluation.

EDUCATIONAL, CULTURAL, AND RECREATIONAL OPPORTUNITIES

It has already been stated that the distribution of income, and in particular its spread to certain disadvantaged groups, is an important social effect. Similarly, the distribution of educational, cultural, and recreational opportunities among regions and socio-economic groups may be an important consideration in evaluating a major project.

For example, a major dam may provide a new recreational resource for fishing, boating, picnicking, and swimming (depending on the final end use of the water). The social benefits of these recreational uses may be as important as the economic benefits of the water provided by the dam. In some cases it may be advantageous to choose a dam site close to a population centre in order to increase the recreational opportunities to a particular group of people.

Educational and cultural values may be enhanced by engineering projects providing improved access to sites of historical, archeological, or scientific interest. For example, a new road along the coast may provide improved access to some unique geological formations which are of considerable interest to the public.

NATIONAL SECURITY AND EMERGENCY PREPAREDNESS

A new road or railway could have important consequences for national security. Such benefits are difficult to quantify in economic terms and, in fact, only become apparent in times of armed conflict. However, they may be an important social consideration in the evaluation of a project.

Emergency preparedness requires, for example, the provision of a flexible water supply, electricity grid, and road and rail network. The provision of a single dam to supply water to a city or a single power station to supply electricity may be economically efficient under normal circumstances but could be disastrous in the event of a catastrophe such as major technical failure, earthquake, hurricane, nuclear explosion, or sabotage. The provision of some redundancy in all engineering systems is a good rule to follow.

DISPLACEMENT AND RELOCATION

Engineering projects in urban areas may involve the displacement and relocation of some houses, commercial, or industrial buildings. A typical example arises from the land acquisition associated with an improved transport link such as a freeway, arterial road, tramway, or railway. It is common practice for the households or firms displaced to be suitably compensated for the loss of their land, home, shop, or factory.

It is often argued that the loss of physical assets is suitably compensated for by the payment of **fair market value**. However, such an argument ignores the fundamental distinction between **value in exchange** and **value in use**. Consider the market for houses of a particular age and quality. If the current market price for such houses is P_o, this may be called the **value in exchange** of the property. Clearly all homeowners who are prepared to sell at this price or less will put their houses on the market. However, as indicated by the supply curve (see the Appendix to Chapter 5) there are many homeowners who are not willing to sell at the price P_o. To these people, the value in use of the house exceeds its value in exchange. If a road authority were to compulsorily acquire such a house and pay compensation equal to P_o, the homeowner would clearly suffer a loss of utility. Full compensation should cover the value in use of the asset and would usually exceed P_o by a significant margin. A similar argument applies to commercial and industrial properties.

The difference between value in use and value in exchange of a house includes an allowance for the stress of moving, disruption of social ties, the cost of finding a new house, and changes in accessibility and travel costs at the new location. In practice such costs are difficult to assess and considerable negotiation may be required with the property owners.

The compensation costs for displaced households and firms are a true opportunity cost to a project and should be included in the economic evaluation, rather than the social evaluation.

NEIGHBOURHOOD DISRUPTION AND INTRUSION

An engineering project may involve significant disruption and/or intrusion into a community. This is particularly true of transportation corridors which can sever a cohesive community or produce intrusive impacts in the form of noise, vibration, air pollution, or aesthetic degradation. A sense of community is often very strong and may be centred on facilities such as schools, shopping centres, the town hall, or parks. Proposals which isolate one part of the community from another may receive considerable opposition from local community groups. Transport corridors should be planned to provide the minimum disruption to communities and ensure that adequate safe crossings such as overpasses or underpasses are provided for pedestrians and cyclists.

The principle of compensation described in the previous section can be applied to neighbourhood disruption and intrusion, but it is very difficult to define the loss of utility for an entire community. Attempts have been made to assess the economic effects of noise due to airports, highways, or railways by assessing the lower property values adjacent to these facilities. If such an approach is taken, these social effects can be included as a cost in the economic evaluation.

6.4 Techniques for environmental and social evaluation

It is very difficult to compare projects in terms of their environmental and social effects because of the many diverse factors involved. This contrasts with an economic comparison in which a single criterion such as NPV (see Section 5.4) gives an unambiguous ranking of projects. A number of techniques have been developed to assist in evaluating environmental and social factors. Two which will be discussed here are the **matrix display technique** and the **environmental index**.

The matrix display technique

The matrix display technique involves the construction of a large matrix which summarizes all of the environmental and social impacts of a proposed project. Often this information is given in a quantitative form. One example of this technique is the **Leopold matrix** developed by Leopold *et al.* (1971).

The Leopold matrix has across its top a list of 100 actions which could cause environmental impact, and down its left-hand side a list of 88 environmental characteristics which could be affected by the actions. Table 6.1 lists

Table 6.1 Project actions in the Leopold matrix (Leopold *et al*. 1971).

A. Modification of regime
(a) Exotic flora or fauna introduction
(b) Biological controls
(c) Modification of habitat
(d) Alteration of ground cover
(e) Alteration of ground water hydrology
(f) Alteration of drainage
(g) River control and flow modification
(h) Canalization
(i) Irrigation
(j) Weather modification
(k) Burning
(l) Surface or paving
(m) Noise and vibration

B. Land transformation & construction
(a) Urbanization
(b) Industrial sites and buildings
(c) Airports
(d) Highways and bridges
(e) Roads and trails
(f) Railroads
(g) Cables and lifts
(h) Transmission lines, pipelines and corridors
(i) Barriers including fencing
(j) Channel dredging and straightening
(k) Channel revetments
(l) Canals
(m) Dams and impoundments
(n) Piers, seawalls, marinas & sea terminals
(o) Offshore structures
(p) Recreational structures
(q) Blasting and drilling
(r) Cut and fill
(s) Tunnels and underground structures

C. Resource extraction
(a) Blasting and drilling
(b) Surface excavation
(c) Subsurface excavation and restoring
(d) Well drilling and fluid removal
(e) Dredging
(f) Clear cutting and other lumbering
(g) Commercial fishing and hunting

D. Processing
(a) Farming
(b) Ranching and grazing
(c) Feed lots
(d) Dairying
(e) Energy generation
(f) Mineral processing
(g) Metallurgical industry
(h) Chemical industry
(i) Textile industry
(j) Automobile and aircraft
(k) Oil refining
(l) Food
(m) Lumbering
(n) Pulp and paper
(o) Product storage

E. Land alteration
(a) Erosion control and terracing
(b) Mine sealing and waste control
(c) Strip mining rehabilitation
(d) Landscaping
(e) Harbour dredging
(f) Marsh fill and drainage

F. Resource renewal
(a) Reforestation
(b) Wildlife stocking and management
(c) Ground water recharge
(d) Fertilization application
(e) Waste recycling

G. Changes in traffic
(a) Railway
(b) Automobile
(c) Trucking
(d) Shipping
(e) Aircraft
(f) River and canal traffic
(g) Pleasure boating
(h) Trails
(i) Cables and lifts
(j) Communication
(k) Pipeline

H. Waste emplacement and treatment
(a) Ocean dumping
(b) Landfill
(c) Emplacement of tailings, spoil and overburden
(d) Underground storage
(e) Junk disposal
(f) Oil well flooding
(g) Deep well emplacement
(h) Cooling water discharge
(i) Municipal waste discharge including spray irrigation
(j) Liquid effluent discharge
(k) Stabilization and oxidation ponds
(l) Septic tanks, commercial & domestic
(m) Stack and exhaust emission
(n) Spent lubricants

I. Chemical treatment
(a) Fertilization
(b) Chemical deicing of highways, etc
(c) Chemical stabilization of soil
(d) Weed control
(e) Insect control (pesticides)

J. Accidents
(a) Explosions
(b) Spills and leaks
(c) Operational failure

Others
(a)
(b)

the project actions while Table 6.2 lists the environmental characteristics. Other actions or environmental characteristics can be added as appropriate to give over 8800 possible interactions. For any particular project only some of the possible interactions will be relevant, but the actions and environmental characteristics do provide a useful checklist to ensure that all possible effects have been considered.

Each cell of the matrix for which an impact is likely to occur is identified and divided by a diagonal line. An assessment is then made of the *magnitude* of the likely impact and a number between 1 and 10 placed in the upper left-hand corner. 1 represents the smallest magnitude and 10 the greatest. This should be a factual assessment which should not include value judgements.

A number between 1 and 10 is placed in the lower right-hand corner of each cell to indicate the *importance* of the impact. This is based on the value judgement of the evaluator.

It should be emphasized that the matrix is only a way of summarizing the environmental impact of a proposed project. It must be supported by an environmental impact statement which discusses all impacts identified in the matrix.

For example, Figure 6.2 shows a reduced impact matrix for a proposed phosphate mining lease in Los Padres National Forest, California (Leopold *et al.* 1971). The mining will be an open-cut operation with ore processing on site including crushing, leaching, and neutralization.

Project actions

Environmental characteristics	Industrial sites and buildings	Highways and bridges	Transmission lines	Blasting and drilling	Surface excavation	Mineral processing	Trucking	Emplacement of taillings	Spills and leaks
Water quality					2/2	1/1		2/2	1/4
Atmospheric quality						2/3			
Erosion		2/2			1/1			2/2	
Deposition, sedimentation		2/2			2/2			2/2	
Shrubs					1/1				
Grasses					1/1				
Aquatic plants					2/2			2/3	1/4
Fish					2/2			2/2	1/4
Camping and hiking					2/4				
Scenic views and vistas	2/3	2/1	2/3		3/3		2/1	3/3	
Wilderness qualities	4/4	4/4	2/2	1/1	3/3	2/5	3/5	3/5	
Rare and unique species		2/5		5/10	2/4	5/10	5/10		
Health and safety							3/3		

Figure 6.2 The reduced matrix for a phosphate mining lease (Leopold *et al.* 1971).

Table 6.2 Environmental characteristics in the Leopold matrix (Leopold *et al.* 1971).

A. Physical and chemical characteristics	
1. Earth	3. Atmosphere
(a) Mineral resources	(a) Quality (gases, particulates)
(b) Construction material	(b) Climate (micro, macro)
(c) Soils	(c) Temperature
(d) Land form	4. Processes
(e) Force fields & background radiation	(a) Floods
(f) Unique physical features	(b) Erosion
2. Water	(c) Deposition (sedimentation, precipitation)
(a) Surface	(d) Solution
(b) Ocean	(e) Sorption (ion exchange, complexing)
(c) Underground	(f) Compaction and settling
(d) Quality	(g) Stability (slides, slumps)
(e) Temperature	(h) Stress-strain (earthquake)
(f) Recharge	(i) Air movements
(g) Snow, ice, and permafrost	
B. Biological conditions	
1. Flora	2. Fauna
(a) Trees	(a) Birds
(b) Shrubs	(b) Land animals including reptiles
(c) Grass	(c) Fish and shellfish
(d) Crops	(d) Benthic organisms
(e) Microflora	(e) Insects
(f) Aquatic plants	(f) Microfauna
(g) Endangered species	(g) Endangered species
(h) Barriers	(h) Barriers
(i) Corridors	(i) Corridors
C. Cultural factors	
1. Land Use	3. Recreation
(a) Wilderness & open spaces	(a) Hunting
(b) Wetlands	(b) Fishing
(c) Forestry	(c) Boating
(d) Grazing	(d) Swimming
(e) Agriculture	(e) Camping and hiking
(f) Residential	(f) Picnicking
(g) Commercial	(g) Resorts
(h) Industrial	4. Cultural status
(i) Mining and quarrying	(a) Cultural patterns (life style)
2. Aesthetics and human interest	(b) Health and safety
(a) Scenic views and vistas	(c) Employment
(b) Wilderness qualities	(d) Population density
(c) Open space qualities	5. Manmade facilities and activities
(d) Landscape design	(a) Structures
(e) Unique physical features	(b) Transportation network (movement, access)
(f) Parks and reserves	(c) Utility networks
(g) Monuments	(d) Waste disposal
(h) Rare & unique species or ecosystems	(e) Corridors
(i) Historical or archaeological sites and objects	(f) Birds
(j) Presence of misfits	
D. Ecological relationships such as:	**Others**
(a) Salinization of water resources	(a)
(b) Eutrophication	(b)
(c) Disease-insect vectors	
(d) Food chains	
(e) Brush encroachment	
(f) Other	

The most important impact of the proposal is its likely effect on the California condor, a rare and endangered species which exists in the region. The primary actions of concern are blasting and the increase in truck traffic, both of which are likely to disturb the nesting of the condor. In addition, sulphur fumes from mineral processing could prevent the birds from landing to catch prey and hence present a danger to them. These effects are shown in the row labelled 'rare and unique species' as having a magnitude of 5 and an importance of 10.

Other effects considered to be of moderate importance include the impact of industrial sites and buildings, highways and bridges, surface excavation, trucking, and the emplacement of tailings on the 'wilderness qualities' of the area.

The values in the boxes of a Leopold matrix are on an ordinal scale and cannot therefore be added or averaged. However, two or more projects can be compared in terms of the entries in an individual cell in the matrix to see which project will have the greatest impact on that particular characteristic.

Environmental index

One difficulty with the matrix display technique is the difficulty in absorbing all of the information contained in the matrix.

In response to this problem, there have been various attempts to develop a single index of environmental quality. Dee *et al.* (1973) present one such approach called an **environmental evaluation system** (EES). The steps involved in this approach are as follows:

- Identify all environmental parameters which are relevant to the range of projects being considered. For water resources projects the list given in Figure 6.1 is a good starting point.
- Choose a suitable measurement scale for each environmental parameter using a cardinal scale where possible. For example, dissolved oxygen levels measured in milligrams per litre, noise levels measured in decibels, the effect on sport fish measured by the likely reduction in the number of fish caught per season, and so on.
- Use expert advice to develop a set of transformations which can be used to convert environmental measurements for each parameter into environmental quality on a scale of 0 to 1. A value of 0 represents the worst possible environmental quality, and 1 the best possible value. Sample transformations from Dee *et al.* are shown in Figure 6.3.
- Relative weights for each environmental variable are chosen using judgement and expert opinion. The numerical weights used in the environmental evaluation system of Dee *et al.* are shown in Figure 6.1.
- Use experts to assess the effect of each alternative project (including 'no project') on each environmental parameter. These are converted into environmental quality (EQ) units using the transformations developed above.

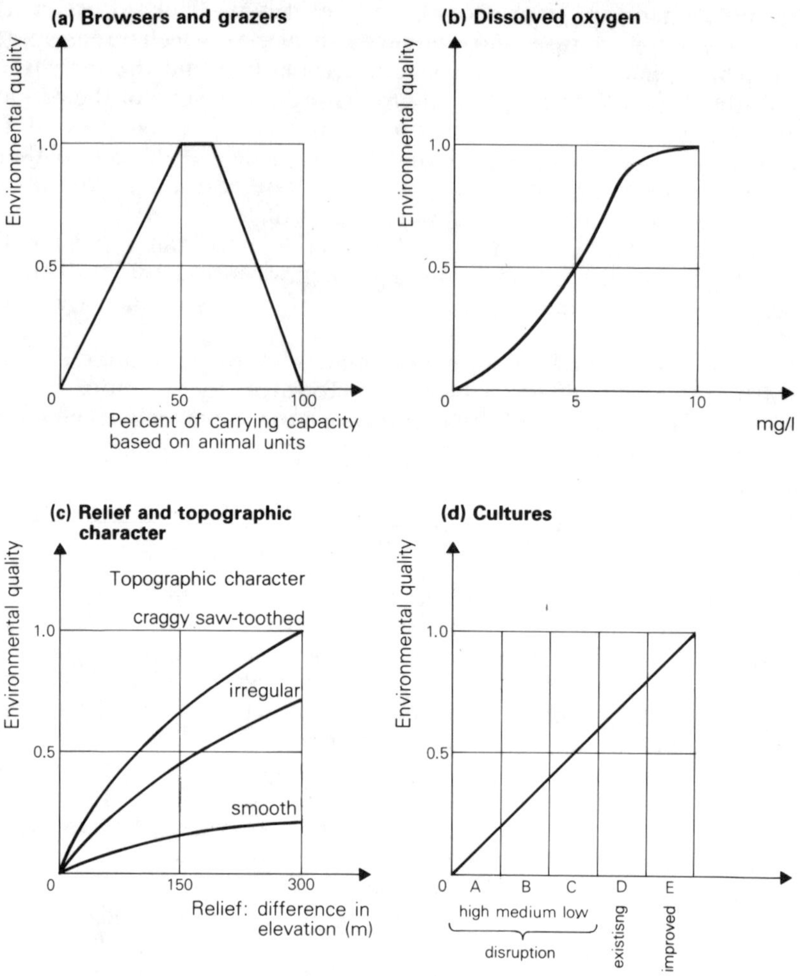

Figure 6.3 Environmental transformations (Dee *et al.* 1973).

- The environmental impact of each project is determined using the following index:

$$\mathrm{EI} = \sum_{i=1}^{m} w_i \left[(V_i)_1 - (V_i)_0 \right] \tag{6.2}$$

where EI is the environmental index, w_i is the relative weight of parameter i, $(V_i)_1$ is the environmental quality of parameter i *with* the project, $(V_i)_0$ is the environmental quality of parameter i *without* the project, and m is the total number of parameters

A high value of EI indicates good environmental quality while a low value means poor environmental quality.

One further feature of the EES is the use of **red flags** to indicate significant changes in one or more environmental parameters. Red flags are also used in areas where there is little or no quantitative data available. Thus, although a particular project may appear to be favourable because of a high value of EI, the existence of one or more red flags would urge caution in specific areas.

This scheme allows projects to be compared in terms of both their economic performance (using NPV) and their effect on environmental quality (using the EI and red flags).

There are, of course, difficulties in any approach which attempts to develop a single index of environmental quality. The major problem is the amount of judgement involved in developing the transformation functions and relative weights for each parameter. Many engineers and environmentalists feel that rather than attempt to incorporate explicit value judgements into the process, it is the role of the analyst to make the maximum amount of information available and understandable to the decision-maker and the public. The use of the impact matrix approach is consistent with this philosophy.

6.5 Multiobjective analysis

In planning and design situations where more than one objective is relevant, it is not usually possible to identify a single 'best' solution. A design which is better than another one in terms of its economic performance might be not as good in terms of its environmental and social impact. A third design which has little environmental impact may involve low economic and social benefits. Compromise is the essence of good planning and design.

The role of the analyst in multiple objective planning is to identify the most efficient designs (in terms of all objectives) and elucidate the tradeoffs among them. A central concept is that of 'inferior' and 'non-inferior' designs or plans.

Suppose only two objectives are relevant in a particular planning problem. These may be national economic development and environmental quality. All feasible designs could be evaluated in terms of these two objectives and plotted in two-dimensional space as shown in Figure 6.4.

By definition, an alternative A is **inferior** to alternative B if B ranks higher or equal to A for all objectives and higher for at least one objective. If A ranks higher than B for one objective and B higher than A for another they are both **non-inferior** (for this pairwise comparison). In Figure 6.4 design X is clearly inferior to design Y and can be discarded from further analysis. Designs Y or Z are non-inferior to each other and should therefore both be retained in the evaluation.

Designs which are not inferior to any other are called 'non-inferior' designs. A line joining all non-inferior designs is called the **transformation curve** (or **transformation surface** if there are more than two objectives). The

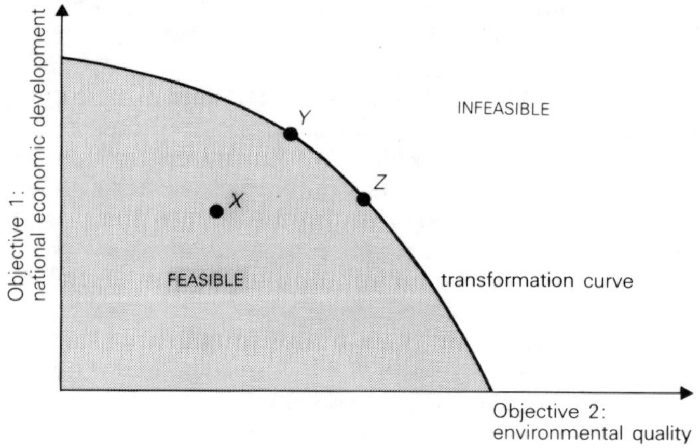

Figure 6.4 Multiobjective tradeoff between alternatives.

transformation curve represents the boundary between feasible and infeasible designs in objective space.

Clearly the preferred design should lie on the transformation curve **provided all relevant objectives have been included** in the analysis. The choice of a final design involves a value judgement as to the relative importance of the objectives and, in many cases, will involve a political decision.

The following case study is presented as an example of multiobjective planning. It is adapted from Major (1974).

The case study concerns the planning of a dam for water supply, recreation and flood control purposes on the Big Walnut Creek in Indiana, USA. The original planning of the dam was carried out by the U.S. Army Corps of Engineers. However, conservationists opposed the design because it involved the partial flooding of a woodland area containing species of trees unique to the region. Subsequently, a team from the Massachusetts Institute of Technology set about applying a multiobjective planning approach to the problem.

Two primary objectives were identified, namely economic development and environmental quality.

The economic development objective was measured in terms of the difference between the benefits due to water supply, recreation and flood control and the costs of construction and maintenance. All benefits and costs were expressed in terms of equivalent annual worth (see Section 5.4).

In this study a single measure of effectiveness was adopted for the environmental quality objective, namely the area of woodland saved (in hectares).

Alternative plans were developed involving dams at various locations and of various heights. The benefits and costs of each design were estimated and these were converted into equivalent annual worth. The maximum area of woodland flooded was computed for each design using a contour map and a knowledge of the maximum water level in the reservoir.

All designs were then presented in terms of the two objectives as shown in Figure 6.5. The boundary of the feasible designs (the transformation curve) is shown. (This is a modified form of the curve given in Major, 1974.) The total area of woodland was approximately 142 ha and hence this was the maximum area which could be saved.

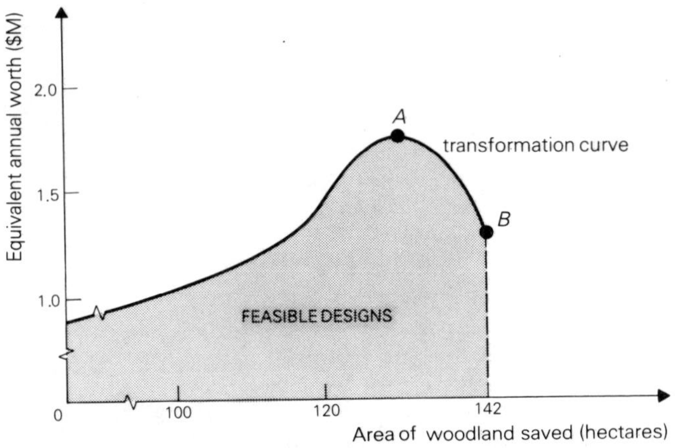

Figure 6.5 Transformation curve for dams on Big Walnut Creek.

Attention was then focused on the non-inferior alternatives represented by the segment AB of the transformation curve. This curve clearly demonstrates the tradeoff between economic benefits and conservation which is implicit in making this type of decision.

The U.S. Army Corps of Engineers at the time wanted to give a high weighting to economic efficiency and so preferred a design close to point A. Conservation groups, on the other hand, put a high priority on preservation of the total integrity of the woodland and so preferred design B. Clearly, there is no single *correct* design and a value judgement is needed in order to reach a decision. Ultimately the design chosen in the Big Walnut study did not involve any significant loss of woodland.

The above approach emphasizes the development of technically efficient (or 'non-inferior') alternatives before the final selection is made. The selection is then based on a value judgement on the part of the decision-maker.

It is interesting to note that the choice of a particular point on the transformation curve implies a relative weight between all objectives. For example, in the case of the Big Walnut scheme, the choice of design C (in Figure 6.6) implies a relative weight at the margin of $0.03M per year per hectare of woodland saved. This is given by the magnitude of the slope of the transformation curve at this point. This information is useful in selecting a final design. For example, a decision-maker who initially selects design C, and then is told the implied weight, may decide that the value of a hectare of woodland saved is somewhat less than \$0.03M per year. This would indicate that the final design should be closer to point A. By interacting with the decision-maker in this way a socially optimal design can be identified.

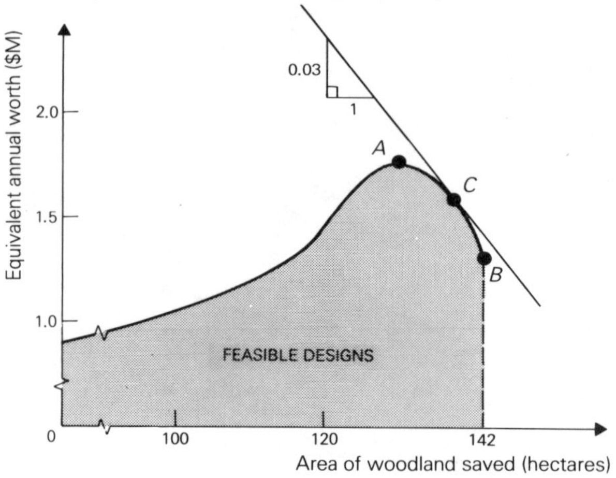

Figure 6.6 Relative weight of objectives for Big Walnut study.

Of course the procedure could be simplified if the decision-maker identifies a set of relative weights on all objectives before any analysis is carried out. This is usually difficult to achieve, as most people find it easier to choose between discrete alternatives than to identify abstract weights.

6.6 Value paths

In the Big Walnut case study only two objectives were deemed to be relevant, and so the display of the transformation curve was relatively simple. In more general cases, there may be many components of the environmental and social objectives and a different form of display is required.

One approach is to use an environmental index and thereby collapse all the environmental variables onto one axis. Similarly an index of social well-being can be developed. Only three dimensions would then be required to identify a transformation surface (i.e. economic, environmental, and social). However, the difficulties in identifying a single environmental index and a single social index detract from this approach.

An alternative approach utilizes **value-paths**. This method will be demonstrated using as an example the problem of finding a water management plan for the Encounter Bay district (Dandy and Tan 1987).

Encounter Bay is a coastal resort area in South Australia with a permanent population of about 10 000. In summer, this population is boosted by a large number of people holidaying in the area. The district may be supplied with water by pumping from three sources as shown in Figure 6.7. These are:

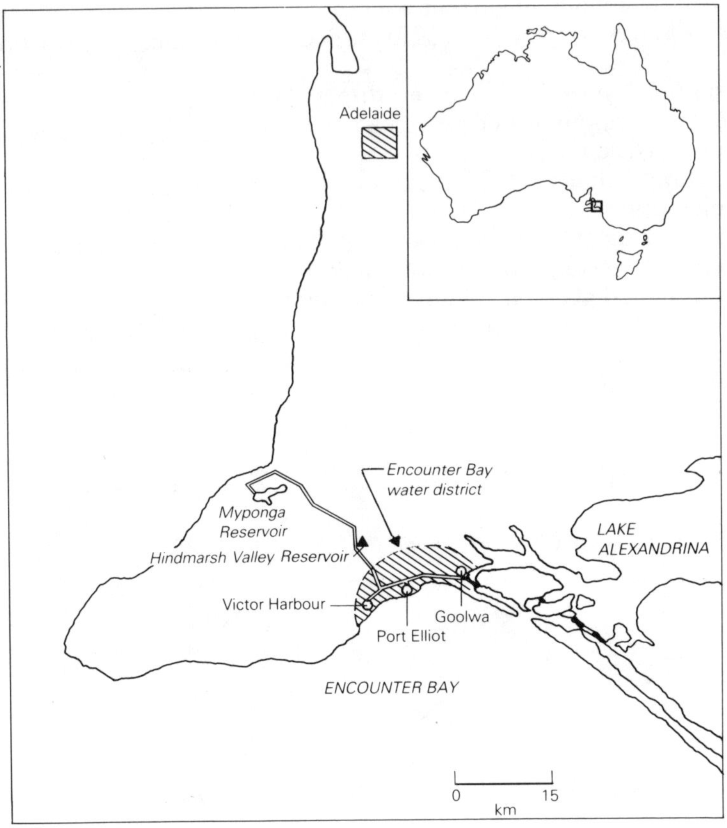

Figure 6.7 The Encounter Bay water supply district.

- Myponga Reservoir
- Hindmarsh Valley Reservoir
- Lake Alexandrina

Myponga Reservoir is of good quality but is the most expensive source because of high pumping cost. Hindmarsh Valley Reservoir periodically has water quality problems in the form of high colour due to organic material from the catchment. Although this is harmless to human health, there is a considerable consumer reaction to colour in the water because it stains laundry and is aesthetically unpleasant. Water in Lake Alexandrina may have high salinity and/or turbidity. At times the levels of these parameters are sufficiently high to cause consumer complaints, but there is no danger to human health. At various times of the year, the volume of water available from each source is limited.

The water authority for this district has a water management plan indicating the priority of use of each source depending on water quality and availability. Dandy and Tan examined alternative policies to see if improvements could be made in current operation.

The measures of effectiveness identified for the water supply were:

- pumping cost evaluated over a three-year period
- the water colour which is exceeded on average in one month out of twelve (Hazen units)
- the water salinity which is exceeded on average in one month out of twelve (mg/l)
- the water turbidity which is exceeded on average in one month out of twelve (nephelometric turbidity units)
- the weighted average colour of the water supplied (Hazen units)
- the weighted average salinity of the water supplied (mg/1)
- the weighted average turbidity of the water supplied (nephelometric turbidity units)

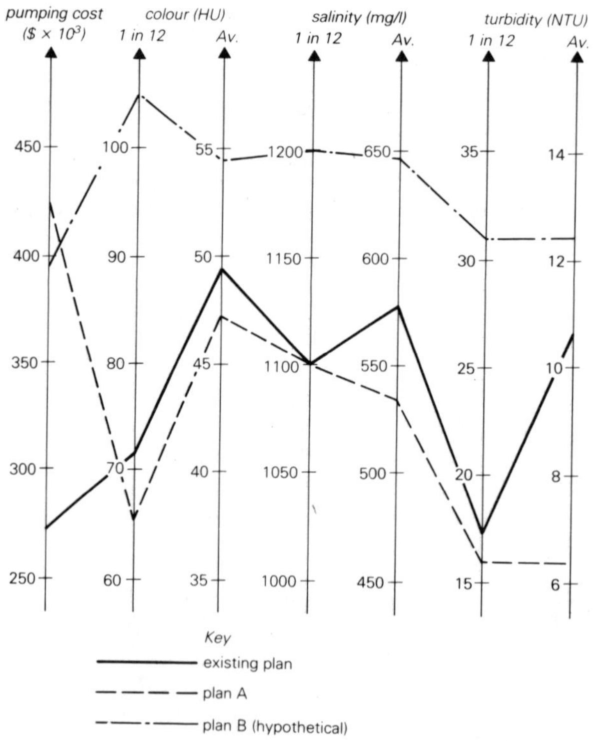

Figure 6.8 Value paths for Encounter Bay water supply district.

For all measures listed above, smaller values are better than larger ones. It should be noted that the water quality measures include average values as well as levels which are exceeded in one month out of twelve. It was considered that both these sets of measures were important in determining public acceptability of the supply.

Some results from the study utilizing the value-path technique are shown in Figure 6.8. Each value-path is for a particular alternative – in this case a particular management plan. It can be seen that Plan A is better than the existing plan for all water quality measures except for the salinity which is exceeded in one month out of twelve (for which it is equal). However, the pumping cost is considerably higher for Plan A. This typifies the type of tradeoff which must be resolved when considering multiple objective problems.

For illustration, a hypothetical plan (Plan B) is plotted in Figure 6.8. Clearly it is inferior to the existing plan, as it is worse according to all measures of effectiveness. Plan B should therefore be discarded from further analysis.

The final choice between the existing plan, Plan A and other non-inferior plans will depend on the community's willingness to pay for improved water quality.

Appendix: Cardinal and ordinal scales of measurement

Every environmental or social parameter has its own scale of measurement. It is important to distinguish between **ordinal** and **cardinal** scales.

An ordinal scale gives an ordering of items according to some criterion. An example is Moh's hardness scale used by geologists. This is given in Table 6.3. As with any ordinal scale the steps between the levels are not necessarily equal, e.g. diamond is about ten times as hard as corundum, which is only 10% harder than topaz. The essential point is that each mineral in the list is harder than the ones below it.

Table 6.3 Moh's hardness scale.

10	Diamond (hardest)
9	Corundum
8	Topaz
7	Quartz
6	Orthoclase
5	Apatite
4	Fluorite
3	Calcite
2	Gypsum
1	Talc (softest)

Another example of an ordinal scale which might be used to rate the visual intrusion of an engineering project is

very high – high – medium – low – very low

Once again one cannot say that a 'very high' intrusion is twice as bad as a 'high' level of intrusion, only that it is worse.

A cardinal scale may be either an **interval** scale or a **ratio** scale. An interval scale has the property that an interval of a specified magnitude represents the same change at any point on the scale. For example, with the Celsius scale for temperature, an interval of 10°C represents the same temperature increase at all points on the scale. This permits averaging to be carried out on scaled values. However, it cannot be said that 100°C represents twice the temperature of 50°C as the zero point of the scale is arbitrary.

Ratio scales have a natural zero point. Examples are mass in kilograms, length in metres, time in seconds, and power in watts. There is no question that a 2 kg body has twice the mass of a 1 kg body.

References

Cohon, J. L. 1978. *Multi-objective programming and planning*. New York: Academic Press.

Dandy, G. C. & K. K. Tan 1987. Use of computer analysis to develop a management plan for an urban water supply system. Proceedings of 2nd International Conference on Education, Practice and Promotion of Computational Methods in Engineering Using Small Computers, Guangzhou, China, 34–37.

Deason, J. P. & K. P. White Jr. 1984. Specification of objectives by group processes in multiobjective water resources planning. *Water Resources Research* **20,** 189–196.

Dee N., J. Baker, N. Drobney, K. Duke, I. Whitman & D. Fahringer 1973. An environmental evaluation system for water resources planning. *Water Resources Research* **9,** 523–535.

deNeufville, R. & J. H. Stafford 1971. *Systems analysis for engineers and managers*. New York: McGraw-Hill.

Finsterbusch, K. 1980. *Understanding social impacts. Assessing the effects of public projects*. Beverly Hills: Sage.

Goicoechea, A., D. R. Hansen & L. Duckstein 1982. *Multi-objective decision analysis with engineering and business applications*. New York: Wiley.

IEA 1987. *Environmental Code of Practice*, Institution of Engineers (Australia). Fyshwick, ACT: Pirie Printers Sales Pty. Ltd.

Leopold, L. B., F. E. Clarke, B. B. Hanshaw & J. R. Balsley 1971. *A Procedure for Evaluating Environmental Impact*. U.S. Geological Survey Circular 645.

Loucks, D. P. & L. Somlyody 1986. Multiobjective assessment of multipurpose water resources projects for developing countries. *Natural Resources Forum* **10,** 61–75.

Major, D. C. 1974. Multiobjective redesign of the Big Walnut Project. In *Systems planning and design: case studies*, deNeufville, R. and D. H. Marks (eds.), 322–37. Englewood Cliffs, NJ: Prentice-Hall.

Mishan, E. J. 1982. *Cost-benefit analysis*, 3rd edn. London: Allen and Unwin.

O'Brien, W. T. 1977. *Multi-objective planning for water resources management – review of recent developments in the USA*. AWRC Technical Paper No. 29, AGPS, Canberra.

O'Brien, W. T., A. A. Thornley & A. S. Atkins 1977. *Multi-objective planning for water resources management*. AWRC Technical Paper No. 27, AGPS, Canberra.

Riggs, J. L. 1977. *Engineering economics*. New York: McGraw-Hill.

USWRC 1973. Principles and standards for planning water and related land resources. United States Water Resources Council. Federal Register, **38** (174), Sept. 10, Washington D.C.

USWRC 1983. Economic and Environmental Principles and Guidelines for Water and Related Land Resources Implementation Studies. United States Water Resources Council. *Federal Register*, March 10, Washington D.C.
Weisbrod, B. A. 1972. Deriving an implicit set of governmental weights for income classes. In *Cost–Benefit Analysis*, R. Layard (ed.), 395–428. London: Penguin.

Problems

6.1 'Project evaluation is simple. All you need to do is put a dollar value on all benefits and costs and undertake all projects which have a benefit–cost ratio greater than 1.' Discuss this statement in the light of:

(a) Environmental and social impacts
(b) The question of how benefits and costs are distributed among groups in the community.
(c) Budget limitations.

6.2 A highway winds its way through a range of hills as shown in Figure 6.9. The hills area consists of natural bushland with an occasional house near the highway. Quite a number of accidents occur on the highway because of its steep grades and tight corners. It is proposed to upgrade the highway by straightening the alignment and reducing the gradient. This will involve the construction of a number of tunnels and/or bridges.

(a) Make a list of the major environmental and social factors which would need to be considered when evaluating alternative routes for the new highway.
(b) How would you display to the general public the environmental and social impacts of alternative routes?

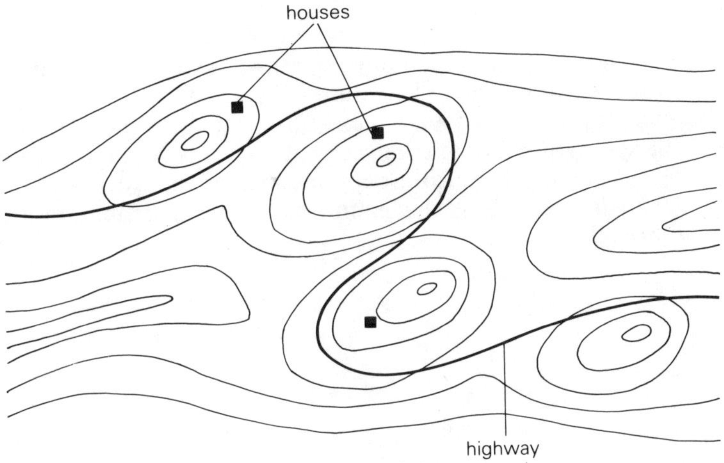

Figure 6.9 Existing highway route.

6.3 A major city needs a new 2000 MW power station to cater for rapidly increasing electricity demands. The station will use natural gas as a fuel. There is an existing gas pipeline from the gasfields which are some 500 km from the city and this pipeline has sufficient capacity to supply the proposed power station (see Figure 6.10). The two alternative sites being considered for the power station are:

A adjacent to the city.
B near the gasfields. (These are in a semi-arid region with a small town and aboriginal reserves nearby.)

Site A has the advantage of being easy to connect into the existing power grid whereas site B would require the construction of a new power transmission line.

What are the economic, environmental, and social tradeoffs involved in choosing between the two sites?

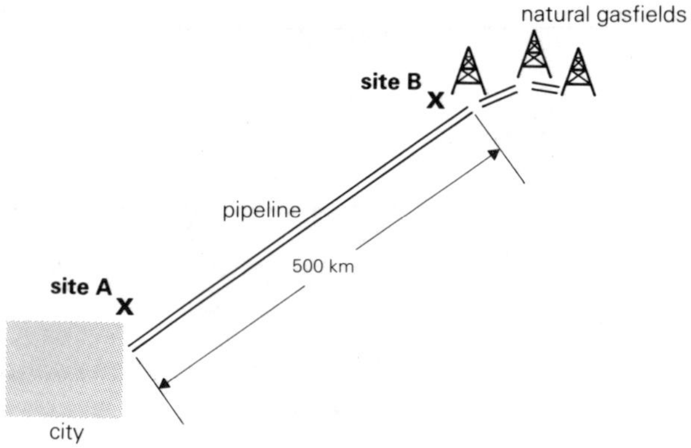

Figure 6.10 Power station sites.

6.4(a) How would you go about developing a single index of water quality for the water supply to Encounter Bay (Section 6.6)? What are the advantages and disadvantages of using such an index?

(b) The state government could set water quality standards which must be met by all town water supplies such as that to Encounter Bay. What do you see as the main advantages and limitations of this approach?

7 Optimization: linear programming

This chapter introduces general concepts of mathematical optimization and then goes on to deal with a specific optimization technique called **linear programming**. *Optimization techniques are applied to problems which can be represented by means of a formal mathematical model consisting of an objective function with various constraints.*

7.1 Introduction

In the processes of planning and design described in Chapter 2 the engineer is attempting to maximize the achievement of a number of specified **objectives** while ensuring that various **constraints** are satisfied. One or more **measures of effectiveness** are used to indicate the degree to which the objectives are achieved. These processes may be thought of in terms of a formal mathematical optimization procedure, although in reality many engineering design problems are too complicated to be formulated in this way. Even if not applied rigorously to a particular problem, optimization theory can provide a useful framework for the activities of planning and design.

In this chapter we first examine the general mathematical optimization problem as it relates to planning and design. We then discuss the technique of linear programming which has been developed to solve a particular class of optimization problems, namely those with a single linear objective function and linear constraints.

7.2 Mathematical optimization

The elements of a mathematical optimization problem are the following:

- A set of n decision variables whose values may be selected by the decision-maker or designer. These are denoted by x_j $(j=1, \ldots, n)$.
- A set of q objective functions which are to be maximized or minimized. Each objective function is a linear or non-linear function of the decision variables. They are denoted by
  $Z_k = g_k(x_1, x_2, \ldots, x_n) \qquad (k=1, \ldots, q)$
- A set of m constraints which limit the feasible combinations of decision variables. They may be linear or non-linear functions of the decision

variables and may be equalities or inequalities of the 'less than or equal to' or 'greater than or equal to' type. They are denoted by
$h_i\,(x_1, x_2, \ldots, x_n)\ \{\leqslant, =, \text{or} \geqslant\}\ b_i \quad (i=1, \ldots, m)$
- A set of known right-hand sides of the constraints denoted by b_i $(i=1, \ldots, m)$. Each one of these may, for example, represent the availability of a certain resource or a minimum standard to be achieved in the design.

The general mathematical optimization problem may then be stated as follows:

Choose values for the n decision variables $(x_1, x_2, \ldots, x_n)$ so as to maximize (or minimize) the q objective functions

$$Z_k = g_k\,(x_1, x_2, \ldots, x_n) \qquad (k=1, \ldots, q) \tag{7.1}$$

subject to the m constraints

$$h_i\,(x_1, x_2, \ldots, x_n)\ \{\leqslant, =, \text{or} \geqslant\}\ b_i \quad (i=1, \ldots, m) \tag{7.2}$$

Example 7.1 An engineer is involved in the preliminary planning of a dam, hydroelectric plant and pipeline to an irrigation area. The layout is shown in Figure 7.1. The basic decision variables for this problem are the capacity of the reservoir $K(\text{m}^3)$, the peak capacity of the hydroelectric plant $P(\text{MW})$ and the capacity of the pipeline to the irrigation area $C(\text{m}^3/\text{s})$.

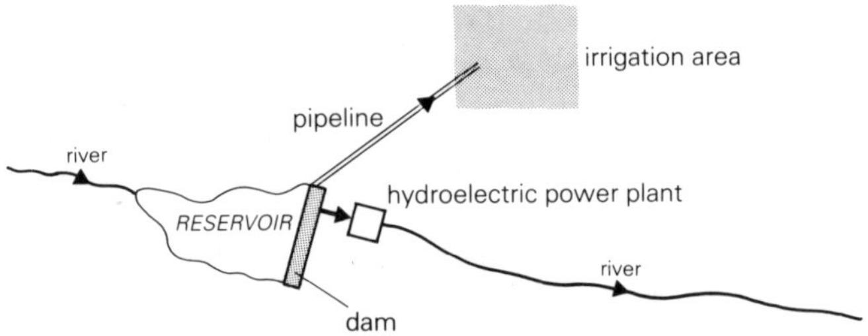

Figure 7.1 A water resources planning problem.

The following are typical objective functions:

(a) To maximize the energy output of the power station, Z_1 (MWh)
(b) To maximize the economic benefits of irrigation, Z_2 (\$)
(c) To minimize the total cost of the scheme, Z_3 (\$)
(d) To minimize the environmental effects of the scheme, Z_4.

In order to state the objective functions and constraints in mathematical terms, we need to define a number of derived variables related to the operation of the scheme:

V_t = the volume of water released through the hydroelectric plant during time period t (m^3) ($t = 1, \ldots, T$)
p_t = the water pressure at turbine level during time period t (kPa) ($t = 1, \ldots, T$)
I_t = the volume of water diverted to irrigation during time period t (m^3) ($t = 1, \ldots, T$)
S_t = the storage in the reservoir at the end of time period t (m^3) ($t = 1, \ldots, T$)
T = the number of time periods considered in the analysis.

The energy output from the scheme during each time period will depend on the product of V_t and p_t, i.e.

$$Z_1 = \sum_{t=1}^{T} \gamma_t V_t p_t h_t$$

where γ_t is a known constant for each time period incorporating the efficiency of the turbines and load peaking effects, among other things, and h_t is number of hours in time period t. The economic benefits of irrigation will depend on I_t, i.e.

$$Z_2 = \sum_{t=1}^{T} f_1(I_t)$$

where $f_i(\cdot)$ denotes a known non-linear function. The total cost of the scheme will depend on K, P, and C, i.e.

$$Z_3 = f_2(K) + f_3(P) + f_4(C).$$

The environmental effects of the scheme will depend principally on K and P, i.e.

$$Z_4 = f_5(K) + f_6(P)$$

The principal constraints will be:

- Continuity of flows at the reservoir:

$$S_t = S_{t-1} + Q_t - V_t - I_t - E_t \; (t = 1, \ldots, T)$$

  where Q_t is the known inflow to the reservoir during time period t (m^3) ($t = 1, \ldots, T$) and E_t is the known loss from the reservoir in time period t (m^3) ($t = 1, \ldots, T$)
- The pressure at the turbines will depend on the storage volume in the reservoir:

$$P_t = f_7(S_t) \qquad (t = 1, \ldots, T)$$

- The rate of energy production of the hydroelectric plant cannot exceed its capacity:

$$\gamma_t V_t p_t \leqslant P \qquad (t = 1, \ldots, T)$$

- The volume of water released to irrigation is limited by the capacity of the pipeline

$$I_t \leqslant C \qquad (t = 1, \ldots, T)$$

- The volume of water stored in the reservoir cannot exceed its capacity:

$$S_t \leqslant K \qquad (t = 1, \ldots, T)$$

- The volume of water released downstream of the dam must be sufficient to support the biota in the river:

$$V_t \geqslant B_t \qquad (t = 1, \ldots, T)$$

  where B_t is the known flow in a time period which is sufficient to support the biota (m^3).
- Non-negativity conditions

$$K, P, C, S_t, I_t, p_t \geqslant 0 \qquad (t = 1, \ldots, T)$$

This is a non-linear optimization problem. Solution techniques for non-linear problems are discussed in Chapter 8.

In an optimization problem with a single objective it is often possible to identify a single optimum solution. However, most planning and design problems, like the one described above, have more than one objective. In this case it is not usually possible to find a single solution which simultaneously achieves the maximum (or minimum) levels of all objectives. The essence of multiple objective optimization problems is to consider tradeoffs between the objectives so as to achieve the **best compromise solution**.

This example clearly demonstrates the conflict among objectives which occurs in many planning and design problems. Maximizing the energy output of the power station would involve diverting all of the available water through the turbines and leaving none for irrigation. This is in conflict with the second objective of maximizing the economic benefits of irrigation. The attainment of both of these objectives requires a large dam, power station, and pipeline which is not consistent with minimizing the total cost of the scheme or its environmental effects.

The number of objectives in the example could be reduced by quantifying the economic benefit of the energy produced and then identifying a single economic objective equal to the benefits of energy plus the benefits of irrigation minus the total cost of the scheme.

The optimization techniques considered in this and the following chapter deal only with problems which have a single objective function. Multi-objective optimization techniques are discussed in Goicoechea *et al*. (1982).

7.3 Linear programming

The simplest case of the general mathematical optimization problem described by conditions (7.1) and (7.2) consists of a single linear objective function and one or more linear constraints. This is called the **linear programming** (LP) problem. It may be written as follows:

$$\text{Max (or Min) } Z = \sum_{j=1}^{n} c_j x_j \tag{7.3}$$

subject to

$$\sum_{j=1}^{n} a_{ij}x_j \; \{\leqslant, =, \text{or} \geqslant\} \; b_i \qquad (i = 1, \ldots, m) \tag{7.4}$$

$$x_j \geqslant 0 \qquad (j = 1, \ldots, n) \tag{7.5}$$

Note that the non-negativity conditions (7.5) are assumed in all LP problems.

Linear programming has been used to solve many large problems in engineering, economics, and commerce. Examples include optimizing the mix of products from a refinery, optimizing the assignment of crews to vehicles for a public transport system, and optimizing the allocation of water among conflicting uses.

Example 7.2 The engineer for the Bullbar City Council must decide how to allocate the Council's workforce and machinery among various activities. For simplicity it will be assumed that only two activities are possible, constructing roads and constructing drains. The engineer has estimated that each kilometre of new road brings a net benefit of \$30 000 per year to the community, whereas each kilometre of drain brings a net benefit of \$20 000 per year. A kilometre of road requires 250 person-days of labour and 640 machine-hours to construct. On the other hand, a kilometre of drain requires 500 person-days of labour and 320 machine-hours to construct. The Council has a workforce of 50 people and 10 machines. Assuming 200 effective working days of 8 hours each per year, write an LP problem to determine the mix of roads and drains which the Council should undertake each year in order to maximize net benefits.

Solution Let x_1 be the number of km of roads to be constructed in a year, and x_2 be the number of km of drains to be constructed in a year.

The objective function to be maximized is the net benefits per year (in thousands of dollars), i.e.

$$\text{Max } Z = 30x_1 + 20x_2$$

Constraints are imposed by the available workforce and machinery. Considering the workforce first, the available resource is 50 people for 200 days, a total of 10 000 person-days. The required resource is $(250x_1 + 500x_2)$. Therefore

$$250x_1 + 500x_2 \leqslant 10\,000$$

Similarly for machinery, the available resource is 10 machines × 200 days × 8 hours/day, i.e. 16 000 machine-hours.
Therefore

$$640x_1 + 320x_2 \leqslant 16\,000$$

In addition, the lengths of roads and drains constructed cannot be negative, and so

$$x_1 \geqslant 0$$
$$x_2 \geqslant 0$$

Therefore the problem formulation becomes

$$\text{Max } Z = 30x_1 + 20x_2 \tag{7.6}$$

subject to:

$$250x_1 + 500x_2 \leqslant 10\,000 \tag{7.7}$$

$$640x_1 + 320x_2 \leqslant 16\,000 \tag{7.8}$$

$$x_1, x_2 \geqslant 0 \tag{7.9}$$

7.4 Graphical solution of LP problems

Before considering the solution of LP problems, some definitions are required.

Any set of decision variables which satisfies all of the constraints is called a **feasible** solution. For a maximization problem, any feasible solution which produces a value of the objective function not less than that produced by any other feasible solution is called an **optimum** solution.

A graphical solution technique will be illustrated for Example 7.2 given above. Firstly, consider constraint (7.7). If this were an equality it could be plotted as a straight line in (x_1, x_2) space as shown in Figure 7.2. All combinations of x_1 and x_2 which satisfy this constraint lie below it. Clearly any solution above this line is infeasible. This is indicated by shading on the infeasible side of the constraint as shown.

In a similar fashion by plotting the other constraints (including the non-negativity constraints) the feasible solution space (or feasible region) can be identified. This is the region ABCD in Figure 7.3. The **feasible solution space** is the set of all combinations of the decision variables which satisfy all of the constraints of the problem.

The optimum solution may be determined by plotting contours of the

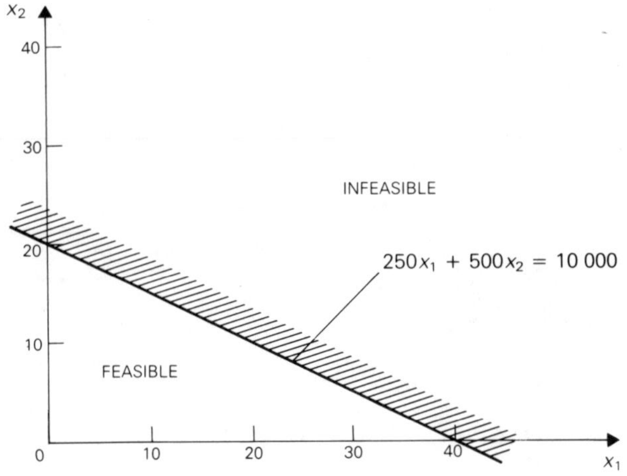

Figure 7.2 Workforce constraint for the Bullbar City Council problem.

objective function. As shown in Figure 7.4 these are parallel straight lines for any two-variable LP problem. It should be apparent that the optimum solution for this problem occurs at point C, i.e. at the intersection of the constraints (7.7) and (7.8). This corresponds to the highest value of Z which can be achieved within the feasible region.

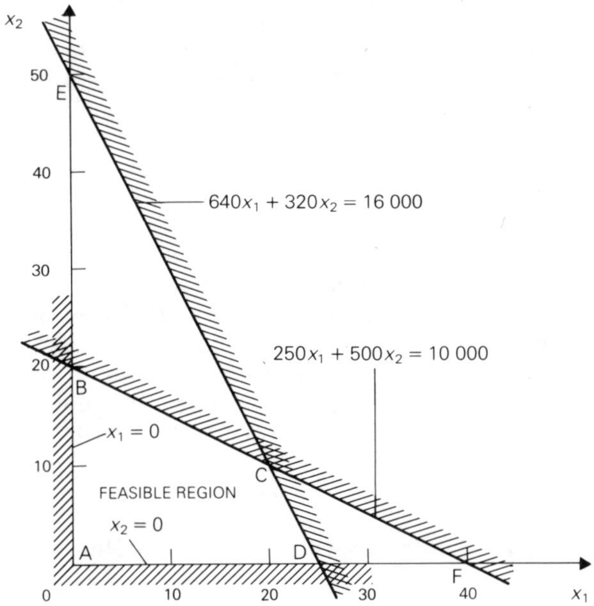

Figure 7.3 Feasible region for the Bullbar City Council problem.

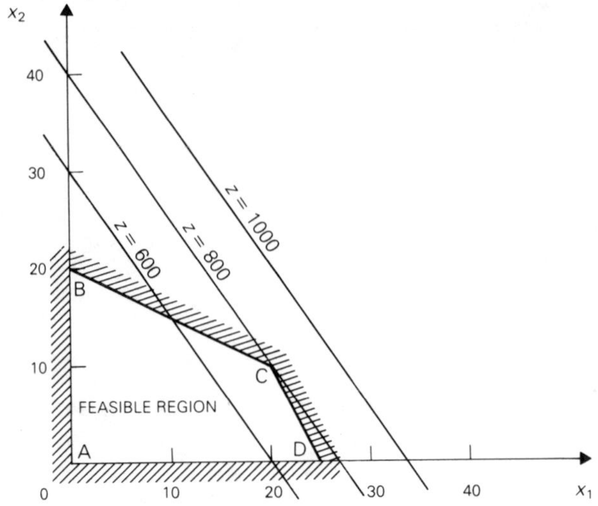

Figure 7.4 Optimum solution to the Bullbar City Council problem.

The solution is

$$x_1 = 20\,\text{km}, x_2 = 10\,\text{km}, Z = \$800 \times 10^3 \text{ per year}$$

If the net benefits of drains relative to roads were changed by a sufficient amount, point B or D would become the optimum solution.

It should be noted that for this example, and indeed all LP problems, the feasible solution space is a **convex region**.

A convex region is one which satisfies the property that a straight line joining any two points in the region only contains points in the region. Any region which does not satisfy this property is called a **non-convex region**.

Examples of convex and non-convex regions in two dimensions are illustrated in Figure 7.5.

It should also be apparent from Figure 7.4 that the solution to any two-dimensional LP problem always occurs at one (or two) **corner points**. A corner point is a feasible solution defined by the intersection of two constraints. It can be seen that if the contours of Z were parallel to BC, both B and C and the points in between would be optimum solutions.

In m dimensions an **extreme point** is a feasible solution which occurs at the intersection of m constraints. It can be shown that the optimum solution to a general LP problem always occurs at one (or more) extreme points.

7.5 Canonical and standard form of an LP problem

It is often useful in the solution of LP problems to reformulate them into either canonical form or standard form.

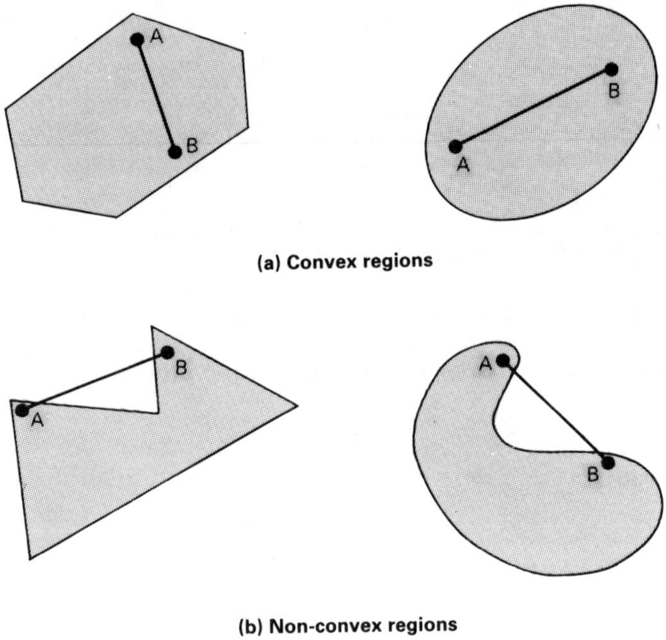

Figure 7.5 Examples of convex and non-convex regions.

The **canonical form** is

$$\text{Max } Z = \sum_{j=1}^{n} c_j x_j \tag{7.10}$$

subject to:

$$\sum_{j=1}^{n} a_{ij} x_j \leqslant b_i \qquad (i = 1, \ldots, m) \tag{7.11}$$

$$x_j \geqslant 0 \qquad (j = 1, \ldots, n) \tag{7.12}$$

It has the following features:

- the objective function is of the *maximization* type
- all constraints are of the 'less than or equal to' type
- the right-hand side constants, b_i, may be positive or negative
- all decision variables are non-negative

Any LP problem may be rearranged into canonical form using one or more of the following operations:

- An objective function of the minimization type can be made into the maximization type by multiplying it by -1, e.g. minimizing $Z = 2x_1 - 3x_2$ is the same as maximizing $Y = -2x_1 + 3x_2$.

- 'Greater than or equal to' constraints can be changed to 'less than or equal to' constraints by multiplying both sides by -1, e.g.
  $5x_1 - 4x_2 \geqslant 12$ is equivalent to $-5x_1 + 4x_2 \leqslant -12$
- Equality constraints can be converted into two inequalities, e.g.
  $4x_1 + 2x_2 = 12$
  is equivalent to
  $4x_1 + 2x_2 \leqslant 12$ and $4x_1 + 2x_2 \geqslant 12$
  i.e.
  $4x_1 + 2x_2 \leqslant 12$ and $-4x_1 - 2x_2 \leqslant -12$
- Variables which are unrestricted in sign may be replaced by two non-negative variables; for example, consider the problem

$$\text{Max } Z = 3y_1 + 2y_2$$

subject to:

$$\begin{aligned} y_1 + y_2 &\leqslant 6 \\ y_1 \quad &\leqslant 8 \\ y_1 \quad &\geqslant 0 \end{aligned}$$

$$y_2 \text{ unrestricted in sign}$$

This may be reformulated by using the substitution $y_2 = (y_2^+ - y_2^-)$ where y_2^+ = the positive part of y_2 and y_2^- = the negative part of y_2. Thus, the problem becomes:

$$\text{Max } Z = 3y_1 + 2y_2^+ - 2y_2^-$$

subject to:

$$\begin{aligned} y_1 + y_2^+ - y_2^- &\leqslant 6 \\ y_1 \qquad\qquad &\leqslant 8 \end{aligned}$$

$$y_1, y_2^+, y_2^- \geqslant 0$$

It can be shown that if y_2^+ is positive in the final solution, y_2^- will be zero and vice versa.

The **standard form** may be written as follows:

$$\text{Max (or Min) } Z = \sum_{j=1}^{n} c_j x_j \qquad (7.13)$$

subject to:

$$\sum_{j=1}^{n} a_{ij}x_j = b_i \qquad (i = 1, \ldots, m) \qquad (7.14)$$

$$x_j \geqslant 0 \qquad (j = 1, \ldots, n) \qquad (7.15)$$

The standard form has the following features:

- the objective function may be of the maximization or minimization type
- all constraints are **equations**, with the exception of the non-negativity conditions which remain as inequalities of the 'greater than equal to' type
- the right-hand side of each constraint must be non-negative
- all variables are non-negative

Any LP problem can be transformed into standard form by using one or more of the following operations (in addition to those listed above):

- Inequality constraints may be converted to equalities by the inclusion of **slack** or **surplus** variables. For example, the 'less than or equal to' constraint,
$$2x_1 + 4x_2 \leqslant 15 \tag{7.16}$$
may be written as
$$2x_1 + 4x_2 + S_1 = 15$$
where S_1 is a **slack variable** which must be non-negative in order to satisfy the original constraint.
Similarly, the 'greater than or equal to' constraint
$$3x_1 + 5x_2 \geqslant 18 \tag{7.17}$$
may be written as
$$3x_1 + 5x_2 - S_2 = 18$$
where S_2 is a **surplus variable** which must be non-negative.
The term slack variable is used because the right-hand side of Equation (7.16) often represents the availability of a particular resource (such as machine-hours in the case of the Bullbar City Council). If S_1 is positive then not all of the available resource is being utilized and there is 'slack' in the solution.
Constraints of the 'greater than or equal to' type often represent minimum requirements. For example, the right-hand side of Equation (7.17) may represent the minimum required output of timber from a particular process. If S_2 is positive the minimum output is being exceeded and there is a 'surplus' output.
- A negative right-hand side can be made positive by multiplying through by -1 (and reversing the sign of the inequality).

7.6 Basic feasible solutions

By examining the standard form of an LP problem it can be seen that the constraints expressed by Equations (7.14) are a set of m linear equations in n unknowns (including slack and surplus variables).

Clearly we would expect n to be greater than m. If n is equal to m there is a unique solution to the constraint set and optimization is not required. If n is less than m, the problem is over-constrained and there is, in general, no solution which satisfies all of the constraint equations.

A **basic** solution to an LP problem is one in which $(n - m)$ of the variables are equal to zero. A **basic feasible** solution to an LP problem is a basic solution in which all variables have non-negative values. Thus a basic feasible solution satisfies the constraints (7.14) and the non-negativity conditions (7.15).

Let us examine Example 7.2 further. In order to identify the basic solutions, we need to express the problem in standard form. This is done by introducing slack variables S_1 and S_2 into constraints (7.7) and (7.8) so that the problem becomes:

$$\text{Max } Z = 30x_1 + 20x_2 \tag{7.18}$$

subject to:

$$250x_1 + 500x_2 + S_1 = 10\,000 \tag{7.19}$$

$$640x_1 + 320x_2 + S_2 = 16\,000 \tag{7.20}$$

$$x_1, x_2, S_1, S_2 \geqslant 0 \tag{7.21}$$

Thus, excluding the non-negativity conditions, there are two constraint equations in four variables (including the slack variables S_1 and S_2). A basic solution may be found by setting any two of the variables equal to zero, and solving Equations (7.19) and (7.20) for the other two. In this manner the six basic solutions shown in Table 7.1 can be identified. Each of these basic solutions is identified by a letter in Figure 7.3. Clearly only the first four of these solutions are basic feasible solutions as the other two do not satisfy the non-negativity conditions (7.21).

Table 7.1 Basic solutions to the Bullbar City Council problem.

Solution	x_1	x_2	S_1	S_2
A	0	0	10 000	16 000
B	0	20	0	9600
C	20	10	0	0
D	25	0	3750	0
E	0	50	−15 000	0
F	40	0	0	−9600

It should be apparent from Figure 7.3 that each basic feasible solution is an **extreme point** of the feasible region. As it has already been shown that the optimum solution to an LP problem is an extreme point it follows that *the optimum solution to an LP problem is always a basic feasible solution.*

Now consider the basic solutions to this problem a little more carefully. A basic solution occurs when two of the variables x_1, x_2, S_1, and S_2 equal zero. If we set S_1 equal to zero in Equation (7.19) we obtain

$$250x_1 + 500x_2 = 10\,000 \qquad \text{(i.e. the line BCF in Figure 7.3)}$$

Similarly, by setting S_2 equal to zero in Equation (7.20) we obtain

$$640x_1 + 320x_2 = 16\,000 \qquad \text{(i.e. the line ECD in Figure 7.3)}$$

The line where $x_1 = 0$ is the x_2 axis, and vice versa.

Thus each constraint line (including the x_1 and x_2 axes) in Figure 7.3 corresponds to one variable being equal to zero. The **basic solutions** (where two variables equal zero) occur at the intersections of two constraint lines. Furthermore each **basic feasible solution** corresponds to the intersection of two constraints within the feasible region. Therefore, the basic feasible solutions must be the extreme points of the feasible region.

7.7 The simplex method

The simplex method is a numerical procedure for solving LP problems. It is based on the property that the optimal solution to an LP problem is always a basic feasible solution. The simplex method uses the following steps:

Step 1 Identify an initial basic feasible solution (i.e. an extreme point).
Step 2 Consider movement from this extreme point to all adjacent extreme points and determine whether it is possible to improve the value of the objective function.
Step 3 If improvement is possible choose the next extreme point such that the maximum rate of change of objective function is achieved and move to this point. Return to Step 2.
Step 4 If no improvement is possible an optimum solution has been obtained.

Recall that at each extreme point $(n - m)$ variables are equal to zero, where n is the total number of variables (including slack and surplus variables) and m is the total number of constraints (excluding the non-negativity constraints).

For any particular basic feasible solution the m variables which are non-zero are called the **basic variables**, while the $(n-m)$ zero variables are called the **non-basic variables**. The basic variables are said to form the **basis** of the solution.

Consider Example 7.2 again. By reference to Table 7.1 it can be seen that at solution A, S_1 and S_2 are the basic variables while x_1 and x_2 are the non-basic variables. A closer examination of Table 7.1 in conjunction with Figure 7.3 reveals that adjacent extreme points (e.g. A and B) have only one variable different in the basis. This property is true for all LP problems. Therefore, the movement from one extreme point to an adjacent one is equivalent to allowing one *non-basic* variable to enter the basis and one *basic* variable to leave. This will be examined in more detail in the following example.

Example 7.3 Solve Example 7.2 using the simplex method.
First, Equations (7.18) to (7.20) are rewritten in the following form:

$$Z - 30x_1 - 20x_2 = 0 \tag{7.22}$$

$$250x_1 + 500x_2 + S_1 = 10\,000 \tag{7.19}$$

$$640x_1 + 320x_2 + S_2 = 16\,000 \tag{7.20}$$

Here all variables (including the objective function, Z) are on the left-hand side and all constants on the right-hand side. It is not necessary to include the non-negativity conditions as all variables will be kept non-negative throughout the solution.

Step 1 Identify an initial basic feasible solution.

This is easy for this problem. If we set $x_1 = x_2 = 0$ we can readily solve Equations (7.22), (7.19), and (7.20) to give $Z = 0$, $S_1 = 10\,000$ and $S_2 = 16\,000$, i.e. the solution point A in Table 7.1 and Figure 7.3.

Step 2 Consider movement from this extreme point to all adjacent extreme points and determine whether it is possible to improve the value of the objective function.

From Figure 7.3 it can be seen that the adjacent extreme points are B and D. Solution B corresponds to bringing x_2 into the basis at the expense of S_1; solution D corresponds to bringing in x_1 at the expense of S_2. The signs of x_1 and x_2 in Equation (7.22) indicate that increasing either variable above zero will increase the value of Z. Therefore improvement is possible.

Step 3 Choose the next extreme point such that the maximum rate of change of objective function is achieved and move to this point.

As stated earlier this is equivalent to deciding which new variable should enter the basis. Upon examining Equation (7.22) we see that the coefficient of x_1 is greater in absolute terms than that of x_2. This coefficient is the rate of change of Z per unit change in the corresponding variable. Clearly x_1 offers the greater rate of increase in Z, so it is chosen to enter the basis.

To move to a new extreme point, we need to determine the maximum value to which x_1 can be increased. From Equation (7.19),

$$x_1 \leqslant 10\,000/250 = 40$$

while from Equation (7.20)

$$x_1 \leqslant 16\,000/640 = 25$$

Clearly Equation (7.20) limits the maximum value of x_1. By increasing x_1 to 25, S_2 will become zero and the solution will have moved to point D in Figure 7.3.

It is useful at this stage to rearrange Equations (7.22), (7.19), and (7.20) so that the basic variables and the objective function, Z, each appear in one equation only. This is achieved by taking the equation which limits the maximum value of the new basic variable, x_1 (i.e. Equation 7.20) and using it to express x_1 in terms of the other variables, i.e.

$$x_1 = 25 - \frac{1}{2}x_2 - \frac{1}{640}S_2$$

Now substituting for x_1 in Equations (7.22) and (7.19) gives

$$Z - 5x_2 + \frac{3}{64}S_2 = 750 \qquad (7.23)$$

$$375x_2 + S_1 - \frac{25}{64}S_2 = 3750 \qquad (7.24)$$

$$x_1 + \frac{1}{2}x_2 + \frac{1}{640}S_2 = 25 \qquad (7.25)$$

As x_2 and S_2 are now non-basic variables (and therefore equal to zero) the current solution is obvious from Equations (7.23) to (7.25). It is $Z = 750$, $S_1 = 3750$, $x_1 = 25$. This corresponds to the solution point D in Table 7.1 and Figure 7.3. Now return to Step 2.

Step 2 It is apparent from Equation (7.23) that introducing x_2 into the basis will increase the value of Z.

Step 3 The maximum value of x_2 from Equation (7.24) is

$$x_2 \leqslant 3750/375 = 10$$

The maximum value from Equation (7.25) is

$$x_2 \leqslant 25/\tfrac{1}{2} = 50$$

As Equation (7.24) limits the increase in x_2, S_1 will become non-basic. We rearrange this equation as follows:

$$x_2 = 10 - \frac{1}{375}S_1 + \frac{1}{960}S_2$$

Substituting into Equations (7.23) and (7.25) yields

$$Z + \frac{1}{75}S_1 + \frac{1}{24}S_2 = 800 \qquad (7.26)$$

$$x_2 + \frac{1}{375} S_1 - \frac{1}{960} S_2 = 10 \tag{7.27}$$

$$x_1 - \frac{1}{750} S_1 + \frac{1}{480} S_2 = 20 \tag{7.28}$$

Thus the new solution is

$$Z = 800, x_2 = 10, x_1 = 20$$

i.e. solution point C in Figure 7.3.
Now return to Step 2.

Step 2 As all the coefficients in Equation (7.26) are positive, no further increase in Z is possible, so the optimum solution has been achieved. Therefore, go to Step 4. An optimum solution has been found.

Thus, in this example the simplex method has examined only three basic feasible solutions. Starting at solution A it moved in the direction of maximum improvement to solution D, then to C. In a larger problem the simplex method evaluates only a small subset of all basic feasible solutions (i.e. extreme points) in finding the optimum.

7.8 The simplex tableau

Most practical-sized LP problems are solved using a computer package based on the simplex method. It is convenient in a computer-based algorithm to operate on a matrix of coefficients or **simplex tableau**.

The development of the simplex tableau will be illustrated for Example 7.3. The initial tableau for this problem is shown in Table 7.2(a). The similarity with Equations (7.22), (7.19), and (7.20) should be noted. It can be seen that the basic variables are listed in the second column from the left. The values of these variables can be found from the right-most column. The body of the tableau contains the coefficients of all variables in the constraint equations including the objective function which is usually the first row. The constraint equations must always be written in a form whereby each basic variable appears in only one equation and has a coefficient of one in that equation.

When applied to a tableau for a maximization problem the simplex method proceeds in the following steps:

Step 1 Examine all the coefficients in the objective function row. If these are all positive or zero no further improvement is possible; the optimum solution has been achieved. Otherwise go to Step 2.

Step 2 Select the variable with the most negative coefficient to enter the basis.

Step 3 Find the maximum value which the entering variable can take by dividing the right-hand side of each row (except the objective function row) by the coefficient of the entering variable in that row (if positive). Select the smallest of these values. The corresponding row is called the **pivot row**.

Step 4 The variable which will leave the basis is the basic variable in the pivot row.

Step 5 The coefficient of the entering variable in the pivot row is called the **pivot element**. Divide through the pivot row by the pivot element.

Step 6 Eliminate the entering variable from all other rows by subtracting a multiple of the pivot row. Return to Step 1.

The steps listed above are for a maximization problem. If the objective function is to be minimized the following changes are made to Steps 1 and 2:

- In Step 1, no improvement is possible if all coefficients in the objective function row are *negative* or zero.
- In Step 2, select the variable with the highest *positive* coefficient to enter the basis.

These steps will be applied to Table 7.2(a).

Step 1 As there are negative coefficients in the objective function row, the optimum has not been achieved – go to Step 2.

Step 2 As x_1 has the most negative coefficient it will enter the basis.

Step 3 Dividing the right-hand side element by the coefficient of x_1 in each row we obtain 10 000/250 = 40 for row (2) and 16 000/640 = 25 for row (3). As row (3) has the smaller value, it becomes the pivot row.

Step 4 The basic variable in row (3) is S_2. This will leave the basis.

Step 5 The pivot element is the coefficient of x_1 in row (3), i.e. 640. Divide through row (3) by 640. The new tableau is given in Table 7.2(b).

Step 6 Eliminate x_1 from row (1) in Table 7.2(b) by adding 30 times row (3A). Similarly subtract 250 times row (3A) from row (2) to eliminate x_1 from the latter row. The modified tableau is given in Table 7.2(c). The correspondence between this tableau and Equations (7.23)–(7.25) should be apparent. Now return to Step 1.

Step 1 As the coefficient of x_2 is negative, further improvement in the objective function is possible.

Step 2 As x_2 is the only variable with a negative coefficient, it will enter the basis.

Step 3 Dividing the right-hand-side element by the coefficient of x_2 in each row we obtain 3750/375 = 10 for row (2A) and 25/½ = 50 for row (3A). As row (2A) has the smaller value, it becomes the pivot row.

Step 4 The basic variable in row (2A), S_1, will leave the basis.

Step 5 The pivot element is the 375 in row (2A). Divide row (2A) by this value to give row (2B) (Table 7.2(d)).

Step 6 Eliminate x_2 from rows (1A) and (3A) by subtracting a multiple of the new row (2B). The new simplex tableau is shown in Table 7.2(d). Again the similarity with Equations (7.26)–(7.28) should be noted.

Table 7.2 Simplex tableaux for Example 7.3.

(a) Initial tableau

Row	Basic variable	Coefficient of: Z	x_1	x_2	S_1	S_2	Right-hand side
(1)	Z	1	−30	−20	0	0	0
(2)	S_1	0	250	500	1	0	10 000
(3)	S_2	0	640	320	0	1	16 000

(b) Tableau after Step 5

Row	Basic variable	Coefficient of: Z	x_1	x_2	S_1	S_2	Right-hand side
(1)	Z	1	−30	−20	0	0	0
(2)	S_1	0	250	500	1	0	10 000
(3A)	x_1	0	1	$\frac{1}{2}$	0	$\frac{1}{640}$	25

(c) Tableau after one iteration

Row	Basic variable	Coefficient of: Z	x_1	x_2	S_1	S_2	Right-hand side
(1A)	Z	1	0	−5	0	$\frac{3}{64}$	750
(2A)	S_1	0	0	375	1	$-\frac{25}{64}$	3750
(3A)	x_1	0	1	$\frac{1}{2}$	0	$\frac{1}{640}$	25

(d) Final tableau (after two iterations)

Row	Basic variable	Coefficient of: Z	x_1	x_2	S_1	S_2	Right-hand side
(1B)	Z	1	0	0	$\frac{1}{75}$	$\frac{1}{24}$	800
(2B)	x_2	0	0	1	$\frac{1}{375}$	$-\frac{1}{960}$	10
(3B)	x_1	0	1	0	$-\frac{1}{750}$	$\frac{1}{480}$	20

Now return to Step 1. There are no negative coefficients in row (1B) so the optimum solution has been identified. The optimum solution may be read from the final simplex tableau as: $Z = 800, x_2 = 10, x_1 = 20, S_1 = S_2 = 0$. This corresponds to the optimum solution identified previously.

7.9 Some difficulties in solving LP problems

Obtaining an initial basic feasible solution

In Example 7.3 it was fortunate that the slack variables S_1 and S_2 appeared in the initial constraint Equations (7.19) and (7.20) in such a way as to form an initial basic feasible solution. In general this will not always be the case. In fact, when the original problem contains 'greater than or equal to' and/or equality constraints an initial basic feasible solution will not be obvious.

This difficulty can be overcome by using the two-phase simplex method. In this method, new variables called **artificial variables** are introduced into all constraints which do not have a basic variable. These artificial variables will be the basic variables in the initial solution. The simplex method is then applied in the following two phases:

Phase 1 A new objective function equal to the sum of all artificial variables is minimized. This should force all artificial variables to become non-basic (i.e. to equal zero).

Phase 2 All artificial variables are eliminated from the tableau and the original objective function is optimized.

If all artificial variables are not non-basic at the end of Phase 1, then there is no feasible solution to the original problem.

Example 7.4 *Use of artificial variables.*

$$\text{Min } Z = 3x_1 + x_2 \tag{7.29}$$

subject to:

$$x_1 + 2x_2 \geqslant 10 \tag{7.30}$$

$$x_1 - x_2 \leqslant 0 \tag{7.31}$$

$$2x_1 + x_2 = 12 \tag{7.32}$$

$$x_1, x_2 \geqslant 0 \tag{7.33}$$

Solution First convert the problem into standard form by introducing a surplus variable into Equation (7.30) and a slack variable into Equation (7.31) as shown:

$$\text{Min } Z = 3x_1 + x_2 \tag{7.29}$$

subject to:

$$x_1 + 2x_2 - S_1 = 10 \tag{7.34}$$

$$x_1 - x_2 + S_2 = 0 \tag{7.35}$$

$$2x_1 + x_2 = 12 \tag{7.32}$$

Note that S_2 can be the basic variable in Equation (7.35) as it does not appear in the other equations. The other constraint equations do not have basic variables. (S_1 is unsuitable in Equation (7.34) because of its negative coefficient.)

Form an initial basic solution by adding artificial variables A_1 and A_2 to Equations (7.34) and (7.32) respectively. The constraints become:

$$x_1 + 2x_2 - S_1 + A_1 = 10 \tag{7.36}$$

$$x_1 - x_2 + S_2 = 0 \tag{7.35}$$

$$2x_1 + x_2 + A_2 = 12 \tag{7.37}$$

Now use the simplex method to minimize $Z' = A_1 + A_2$ subject to constraints (7.36), (7.35), and (7.37). This should result in the artificial variables A_1 and A_2 becoming zero. Firstly, the objective function Z' must be expressed in terms of the non-basic variables only. Thus substituting for A_1 and A_2 using Equations (7.36) and (7.37) the Phase 1 objective function becomes:

$$\text{Min } Z' = -3x_1 - 3x_2 + S_1 + 22 \tag{7.38}$$

The initial tableau for the Phase 1 problem is shown as Table 7.3(a). The final simplex tableau after three iterations is shown as Table 7.3(b). The solution at this point is $x_1 = 4$, $x_2 = 4$, and $S_1 = 2$ which can be shown to be feasible by checking Equations (7.34), (7.35) and (7.32). As all artificial variables have left the basis this is the end of Phase 1.

To commence Phase 2, drop all artificial variables out of the tableau and replace the objective function row with the original objective function. Again this must be expressed in terms of the non-basic variables (in this case only S_2 is non-basic in the current tableau).
The original objective function is

$$\text{Min } Z = 3x_1 + x_2 \tag{7.29}$$

From Table 7.3(b),

$$x_2 - \tfrac{2}{3} S_2 = 4 \tag{7.39}$$

$$x_1 + \tfrac{1}{3} S_2 = 4 \tag{7.40}$$

substituting for x_1 and x_2 into Equation (7.29) gives

$$\text{Min } Z = -\tfrac{1}{3} S_2 + 16 \tag{7.41}$$

Therefore, the initial tableau for Phase 2 is shown in Table 7.3(c). The optimum solution is found after one more iteration. The final simplex tableau for Phase 2 is shown in Table 7.3(d).
The optimum solution is

$$Z = 12, x_1 = 0, x_2 = 12$$

Table 7.3 Use of artificial variables.

(a) Initial tableau for Phase 1

Basic variable	Coefficient of: Z'	x_1	x_2	S_1	S_2	A_1	A_2	Right-hand side
Z'	1	3	3	−1	0	0	0	22
A_1	0	1	2	−1	0	1	0	10
S_2	0	1	−1	0	1	0	0	0
A_2	0	2	1	0	0	0	1	12

(b) Tableau at the end of Phase 1

Basic variable	Coefficient of: Z'	x_1	x_2	S_1	S_2	A_1	A_2	Right-hand side
Z'	1	0	0	0	0	−1	−1	0
x_2	0	0	1	0	−⅔	0	⅓	4
x_1	0	1	0	0	⅓	0	⅓	4
S_1	0	0	0	1	−1	−1	1	2

(c) Initial tableau for Phase 2

Basic variable	Coefficient of: Z	x_1	x_2	S_1	S_2	Right-hand side
Z	1	0	0	0	⅓	16
x_2	0	0	1	0	−⅔	4
x_1	0	1	0	0	⅓	4
S_1	0	0	0	1	−1	2

(d) Final tableau for Phase 2

Basic variable	Coefficient of: Z	x_1	x_2	S_1	S_2	Right-hand side
Z	1	−1	0	0	0	12
x_2	0	2	1	0	0	12
S_2	0	3	0	0	1	12
S_1	0	3	0	1	0	14

Most computer packages for solving LP problems automatically include artificial variables to obtain an initial basic feasible solution.

Degenerate solutions

A degenerate solution occurs when one or more of the basic variables is equal to zero. Table 7.3(a) contains a degenerate solution as $S_2 = 0$. One degenerate solution may be followed by another degenerate solution in which no improvement in the objective function occurs. This does not cause a problem with the simplex method and it is quite possible that several degenerate solutions may be followed by non-degenerate solutions in which an improvement in objective function is obtained. On some occasions, cycling between several degenerate solutions may occur. This rarely happens in practice and most commercial computer packages can identify it.

Unbounded solutions

It is sometimes possible to increase the value of the objective function without limit in an LP problem. This indicates a poorly formulated problem. For example:

$$\text{Max } Z = x_1 + 2x_2$$

subject to:

$$2x_1 + x_2 \geqslant 12$$

$$x_1 + x_2 \geqslant 8$$

$$x_1, x_2 \geqslant 0$$

The feasible region for this problem is shown in Figure 7.6. Clearly the

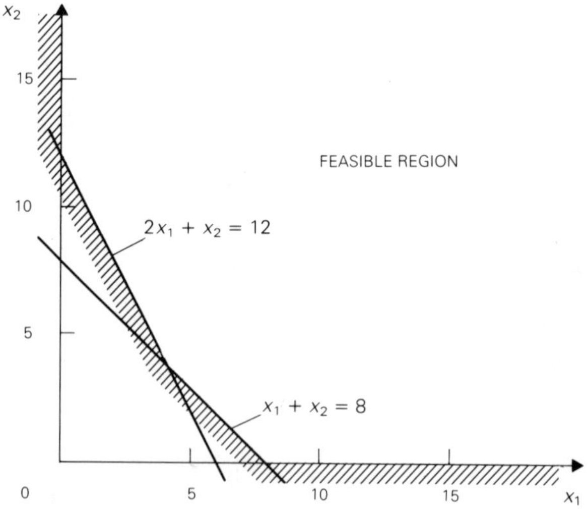

Figure 7.6 Feasible region for an unbounded problem.

objective function may be increased without limit in this feasible region. If the objective function were of the minimization rather than maximization type, an optimum solution could be found (assuming positive coefficients in the objective function).

No feasible solution

Another problem which may arise due to poor problem formulation is that there may be no solution which satisfies all of the constraints. For example:

$$\text{Max } Z = 2x_1 + x_2$$

subject to:

$$2x_1 + 3x_2 \geqslant 30$$

$$2x_1 + x_2 \leqslant 8$$

$$x_1 + 3x_2 \leqslant 9$$

$$x_1, x_2 \geqslant 0$$

The constraints to this problem are shown in Figure 7.7. Clearly there is no feasible region and hence no feasible solution to this problem.

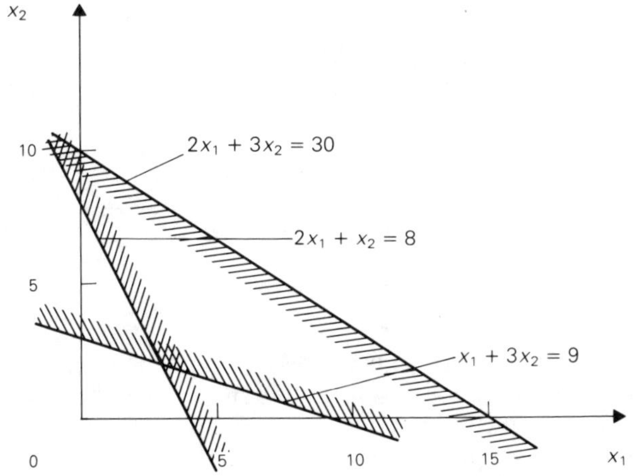

Figure 7.7 No feasible solution.

Alternative optima

Occasionally an LP problem has more than one optimum solution. This occurs when the objective function is parallel to a boundary of the feasible region. For example:

$$\text{Max } Z = 2x_1 + x_2$$

subject to:

$$x_1 + x_2 \leqslant 8$$

$$2x_1 + x_2 \leqslant 12$$

$$x_1, x_2 \geqslant 0$$

The feasible solution space for this problem is shown in Figure 7.8. Also shown is the contour for $Z = 18$. Because contours of the objective function are parallel to the second constraint for this problem, it should be clear that the value of the objective function will be the same at points C and D. So both C and D and all points on the boundary between them are alternative optimum solutions to this problem.

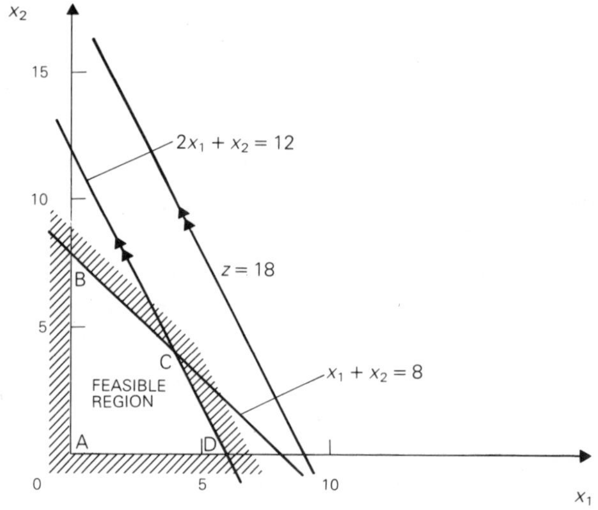

Figure 7.8 Alternative optima.

The simplex method will identify either C or D as the optimum solution and then stop. The existence of alternative optima will be indicated by one or more non-basic variables having a zero coefficient in the objective function row of the final simplex tableau.

Redundant constraints

In some LP problems not all of the constraints are boundaries of the feasible region. Constraints which are not boundaries are called **redundant constraints** because they have no effect on the solution. For example:

$$\text{Max } Z = 2x_1 + x_2$$

subject to:

$$x_1 + x_2 \leqslant 8$$

$$x_1 + 2x_2 \leqslant 12$$

$$6x_1 + 5x_2 \leqslant 60$$

$$x_1, x_2 \geqslant 0$$

The constraints and feasible region for this problem are shown in Figure 7.9. Clearly the constraint $6x_1 + 5x_2 \leqslant 60$ is a redundant constraint as it has no effect on the feasible region.

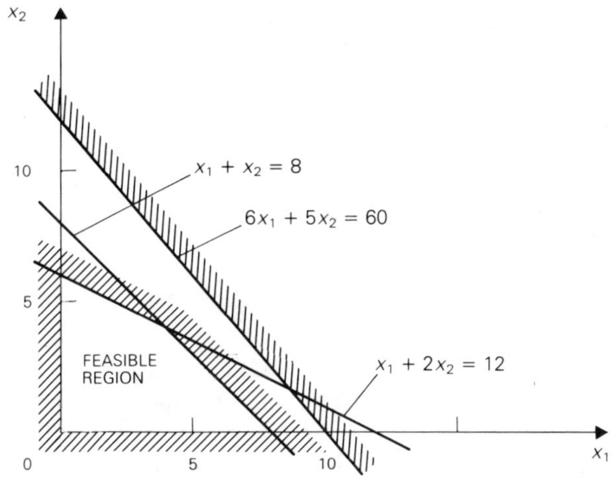

Figure 7.9 Redundant constraint.

The presence of redundant constraints does not in any way inhibit the simplex method from finding an optimum solution. However, because their presence affects the size of the problem and hence the computer time for its solution it is preferable to remove redundant constraints from the formulation if they can be readily identified.

7.10 Examples of LP problems

LP problems occur in many engineering design and planning problems. Some examples are outlined below.

Reservoir design

A common problem in water supply engineering is to determine the required size of a reservoir to provide a specified set of releases to a town or irrigation area. Consider the proposed reservoir shown schematically in Figure 7.10. The design variable is the capacity of the proposed reservoir, K. Flow data have been collected at the proposed site for the last T years. Q_t ($t = 1, \ldots, T$) is the known flow in year t. The known annual water demand in the city is R. E_t ($t = 1, \ldots, T$) represents the annual evaporation and other losses from the reservoir in year t and is assumed to be known. S_t ($t = 0, \ldots, T$) represents the volume of storage in the reservoir at the end of year t. O_t ($t = 1, \ldots, T$) is the spill from the reservoir in year t. S_t ($t = 0, \ldots, T$) and O_t ($t = 1, \ldots, T$) are derived variables whose values will be determined in the course of the optimization.

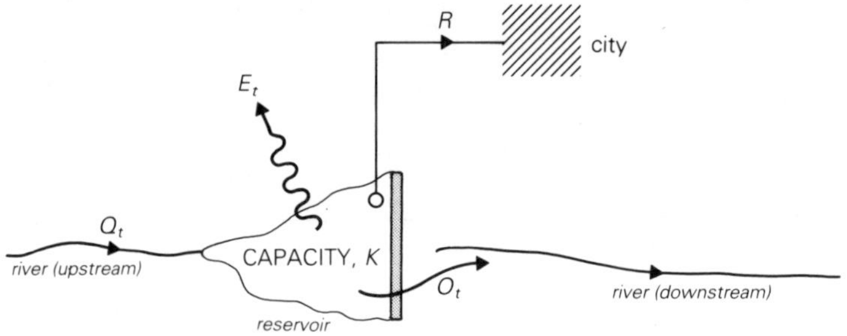

Figure 7.10 Reservoir design problem.

The reservoir size will be determined on the assumption that the historical series of flows will be repeated. The annual demand of the city must be met in every year without the reservoir becoming empty. The objective is therefore to minimize the required capacity, K, subject to the constraints imposed by

- continuity
- the reservoir capacity not being exceeded.

The continuity constraint for year t is as follows:

$$S_t - S_{t-1} = Q_t - E_t - R - O_t$$

i.e. the change in storage in the reservoir in year t equals the total inflow minus the total outflow.

The constraint on capacity at the end of year t is simply

$$S_t \leqslant K$$

It is usual to assume that the reservoir has the same storage at the end of the T years as at the beginning. This allows demand to be met if the historical sequence repeats itself indefinitely, i.e.

$$S_T = S_o$$

Therefore, the LP problem is:

$$\text{Min } Z = K \tag{7.42}$$

subject to:
$$S_t - S_{t-1} + O_t = Q_t - E_t - R \quad (t = 1, \ldots, T) \tag{7.43}$$

$$S_t - K \leqslant 0 \quad (t = 0, \ldots, T) \tag{7.44}$$

$$S_T - S_o = 0 \tag{7.45}$$

$$S_t, O_t, K \geqslant 0$$

where decision variables have been placed on the left-hand side of constraints and known values on the right-hand side.

Road network analysis (adapted from Stark and Nicholls 1972, Example 4–5)

A common problem in transportation planning is the allocation of traffic volumes to a road network. A simple road network is shown in Figure 7.11. The usual demand information for the network will be expressed in terms of an origin–destination matrix, i.e. the volume of traffic wishing to travel from each origin to each designation in a defined time period. Nodes 1 and 3 in Figure 7.11 are origins and nodes 7 and 9 destinations. A typical origin–destination matrix is given in Table 7.4.

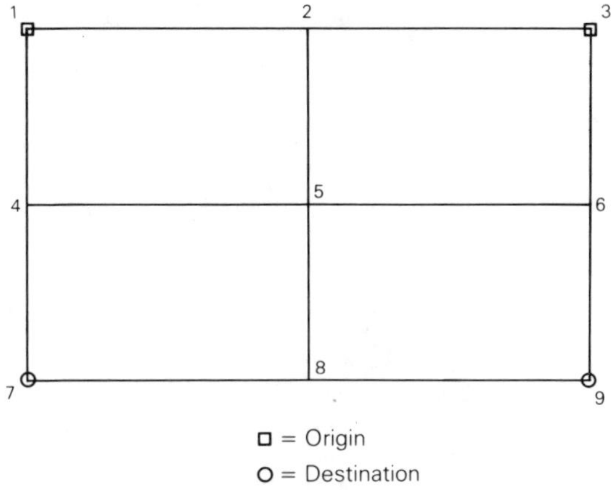

Figure 7.11 Portion of a road network.

Other information available is the capacity (vehicles/h) and the travel time in each direction (minutes) for each road. Typical values are given in Table 7.5. In this example the capacity and travel time is the same in both directions for each road, although this is not necessarily always the case. The objective is to minimize the total time of travel for all traffic.

Table 7.4 Origin–Destination matrix (vehicles/h).

Origin node	Destination Node 7	Destination Node 9
1	4000	3000
3	3000	2000

Let $x_{ij}^{(k)}$ be the volume of traffic (vehicles/h) which originates from node k and travels on road ij, and let K be the set of all origin nodes.

For all traffic originating at node k the total travel time is given by

$$Z^{(k)} = \sum_{i=1}^{n} \sum_{j=1}^{n} \delta_{ij}\, t_{ij}\, x_{ij}^{(k)} \qquad (7.46)$$

where

$$\delta_{ij} = \begin{cases} 1 \text{ if there is a road from node } i \text{ to node } j \\ 0 \text{ otherwise} \end{cases}$$

t_{ij} is the travel time (minutes) on road ij and n is the total number of nodes.

The constraints represent continuity at each node, i.e.

$$\sum_{i=1}^{n} \delta_{ip}\, x_{ip}^{(k)} - \sum_{j=1}^{n} \delta_{pj}\, x_{pj}^{(k)} = D_p^{(k)} \; (p=1, \ldots, n) \qquad (7.47)$$

Table 7.5 Capacities and travel times for road network problem (these are the same in both directions).

Road	Capacity (vehicles/h)	Travel time (minutes)
12	4000	10
23	3000	12
14	4000	15
25	6000	13
36	6000	16
45	2500	8
56	4000	6
47	3500	12
58	5000	10
69	6000	5
78	6000	8
89	2500	7

where

$$D_p^{(k)} = \begin{cases} \text{the volume of traffic which has origin } k \text{ and} \\ \quad \text{destination } p \text{ (for } p \neq k) \\ \text{minus the volume of traffic which originates at node } k \\ \quad (\text{for } p = k) \end{cases}$$

When considering all origin nodes the objective function becomes

$$\text{Min } Z = \sum_{k \in K} \sum_{i=1}^{n} \sum_{j=1}^{n} \delta_{ij}\, t_{ij}\, x_{ij}^{(k)} \tag{7.48}$$

where the first summation is over all origin nodes.
The continuity constraints (7.47) become

$$\sum_{i=1}^{n} \delta_{ip}\, x_{ip}^{(k)} - \sum_{j=1}^{n} \delta_{pj}\, x_{pj}^{(k)} = D_p^{(k)} \qquad \begin{matrix} (p=1, \ldots, n) \\ (k \in K) \end{matrix} \tag{7.49}$$

In addition, capacity constraints for each road must be considered. These are

$$\sum_{k \in K} x_{ij}^{(k)} \leqslant C_{ij} \qquad \text{(for all } i,j \text{ which are connected by a road)} \tag{7.50}$$

where C_{ij} = the capacity of road ij. In addition,

$$x_{ij}^{(k)} \geqslant 0 \qquad (\text{all } i,j,k) \tag{7.51}$$

The LP problem to be solved is defined by the objective function (7.48) and the constraints (7.49), (7.50), and (7.51).

In a more realistic case the travel time on each road depends on the volume of traffic on it. In addition, each driver tries to minimize his own travel time instead of the system being at a global minimum. In these cases, the objective function becomes non-linear but the constraints have the same form.

7.11 Duality

Duality is an important concept in mathematical analysis which often gives valuable insights into the physical problem being solved.

In LP, every problem has another problem associated with it called the **dual**. By solving the original or **primal** problem using the simplex method, the solution to the dual is found automatically. The dual variables have an important interpretation which adds considerable information to an LP solution.

Writing the dual problem

In Section 7.5 the canonical form of an LP problem was given as

$$\text{Max } Z = \sum_{j=1}^{n} c_j x_j \tag{7.10}$$

subject to

$$\sum_{j=1}^{n} a_{ij}x_j \leqslant b_i \qquad (i = 1, \ldots, m) \tag{7.11}$$

$$x_j \geqslant 0 \qquad (j = 1, \ldots, n) \tag{7.12}$$

For an LP problem in canonical form the dual problem may be written using the following rules:

- There is a dual variable corresponding to each primal constraint and a primal variable corresponding to each dual constraint.
- If the primal problem is of the maximization type, the dual problem is of the minimization type and vice versa.
- All of the constraints in the maximization problem are of the 'less than or equal to' type. All of the constraints in the minimization problem are of the 'greater than or equal to' type.
- The coefficients of the objective function in the primal problem are the right-hand sides in the dual problem and vice versa.
- The variables in both problems are non-negative.

To write the dual to the problem given by (7.10)–(7.12), we need to define m dual variables $y_i (i = 1, \ldots, m)$ one corresponding to each primal constraint. The dual problem is:

$$\text{Min} \quad W = \sum_{i=1}^{m} b_i y_i \tag{7.52}$$

subject to

$$\sum_{i=1}^{m} a_{ij}\, y_i \geqslant c_j \qquad (j = 1, \ldots, n) \tag{7.53}$$

$$y_i \geqslant 0 \qquad (i = 1, \ldots, m) \tag{7.54}$$

Note that if these rules are applied to the problem defined by Conditions (7.52)–(7.54), the 'primal' problem is regained (i.e. Conditions (7.10)–(7.12)). Clearly, a symmetry exists between the two problems.

It can be shown that, for any feasible solutions of the primal and dual, $Z \leqslant W$. That is, the objective function value for a feasible solution of the maximization problem is a lower bound on the objective function value of the minimization problem. Furthermore, if Z^* is the maximum value of Z and W^* is the minimum value of W then

$$Z^* = W^* \tag{7.55}$$

i.e. at optimality, both problems have the same objective function value. For example, the primal problem of Example 7.2 is given by (7.6)–(7.9):

$$\text{Max } Z = 30x_1 + 20x_2 \tag{7.6}$$

subject to:

$$250x_1 + 500x_2 \leqslant 10\,000 \tag{7.7}$$

$$640x_1 + 320x_2 \leqslant 16\,000 \tag{7.8}$$

$$x_1, x_2 \geqslant 0 \tag{7.9}$$

As this is already in canonical form, the dual can readily be written as

$$\text{Min } W = 10\,000y_1 + 16\,000y_2 \tag{7.56}$$

subject to:

$$250y_1 + 640y_2 \geqslant 30 \tag{7.57}$$

$$500y_1 + 320y_2 \geqslant 20 \tag{7.58}$$

$$y_1, y_2 \geqslant 0 \tag{7.59}$$

Writing each problem in standard form we obtain for the primal problem:

$$\text{Max } Z = 30x_1 + 20x_2 \tag{7.60}$$

subject to:

$$250x_1 + 500x_2 + S_1 = 10\,000 \tag{7.61}$$

$$640x_1 + 320x_2 + S_2 = 16\,000 \tag{7.62}$$

$$x_1, x_2, S_1, S_2 \geqslant 0 \tag{7.63}$$

and, for the dual problem:

$$\text{Min } W = 10\,000y_1 + 16\,000y_2 \tag{7.64}$$

subject to:

$$250y_1 + 640y_2 - S_3 = 30 \tag{7.65}$$

$$500y_1 + 320y_2 - S_4 = 20 \tag{7.66}$$

$$y_1, y_2, S_3, S_4 \geqslant 0 \tag{7.67}$$

The final simplex tableau for the primal problem is given in Table 7.2(d). It is reproduced here as Table 7.6. The solution is:

Table 7.6 The final simplex tableau for the primal problem.

Basic variable	Coefficient of: Z	x_1	x_2	S_1	S_2	Right-hand side
Z	1	0	0	$1/75$	$1/24$	800
x_2	0	0	1	$1/375$	$-1/960$	10
x_1	0	1	0	$-1/750$	$1/480$	20

$$Z = 800,\ x_1 = 20,\ x_2 = 10,\ S_1 = S_2 = 0$$

The dual problem may be solved using the two-phase simplex method. The final (optimum) tableau is shown in Table 7.7. A_3 and A_4 are the artificial variables corresponding to constraints (7.65) and (7.66) respectively. The solution is:

$$W = 800,\ y_1 = 1/75,\ y_2 = 1/24,\ S_3 = S_4 = 0$$

Table 7.7 The final simplex tableau for the dual problem.

Basic variable	Coefficient of: W	y_1	y_2	S_3	S_4	A_3	A_4	Right-hand side
W	1	0	0	-20	-10	20	10	800
y_2	0	0	1	$-1/480$	$1/960$	$1/480$	$-1/960$	$1/24$
y_1	0	1	0	$1/750$	$-1/375$	$-1/750$	$1/375$	$1/75$

Comparison of Tables 7.6 and 7.7 shows that the optimum values of the dual variables, y_1 and y_2, are given by the coefficients of the slack variables, S_1 and S_2, in the objective function row of the final primal tableau. Similarly, the optimum values of the primal variables, x_1 and x_2, are given by the coefficients of the artificial variables, A_3 and A_4, in the final dual tableau.

In general, the optimum value of each dual variable is given by the final value of the objective function coefficient of the initial basic variable for the corresponding primal constraint. This also applies to the optimum values of the primal variables in relation to the dual constraints. Clearly then the solution to the dual is obtained by solving the primal (and vice versa).

Interpretation of the dual variables

Refer back to the primal problem defined by (7.10)–(7.12). In many problems, the right-hand side element of the ith primal constraint, b_i, may be interpreted as the availability of the ith resource. Combining Equations (7.52) and (7.55), we have at optimality:

$$Z^* = W^* = \sum_{i=1}^{m} b_i y_i^* \qquad (7.68)$$

where y_i^* is the value of y_i at optimality ($i = 1, \ldots, m$).

Therefore, the optimum value of the ith dual variable, y_i^*, may be interpreted as the additional output Z which could be obtained by a unit increase in the ith resource. Thus, if b_i were increased by one unit, we would expect Z^* to increase by y_i^* units.

In the above problem, the dual variable y_1 represents the additional net benefits (in thousands of dollars) if the right-hand side of constraint (7.7) were increased by one unit. In this case this represents an additional person-day of labour. Similarly y_2 represents the additional net benefits (in thousands of dollars) if the right-hand side of constraint (7.8) were increased by one unit, i.e. by having one extra machine-hour available. Therefore

$$y_1 = 1/75 \text{ thousand dollars/person day}$$

$$y_2 = 1/24 \text{ thousand dollars/machine hour}$$

and so

$$y_1 = \$13.33\text{/person-day}$$

$$y_2 = \$41.67\text{/machine-hour}$$

Expressed in this form the dual variables are **shadow prices** for the corresponding resources. The shadow price of a resource is the economic value of having an additional unit of the resource available.

If additional labour can be hired at less than the shadow price of \$13.33 per day the Council would make a net gain by doing so. Similarly if additional machine time can be obtained at less than its shadow price of \$41.67 per hour a net gain can be made. Of course, this gain by employing additional labour or machines will only apply for increases which are not so large as to cause a change in the optimum solution.

References

Beightler, C., D. Phillips & D. Wilde 1979. *Foundations of optimization*, 2nd edn. Englewood Cliffs, NJ: Prentice-Hall.

deNeufville, R. & J. H. Stafford 1971. *Systems analysis for engineers and managers*. New York: McGraw-Hill.

Goicoechea, A., D. R. Hansen & L. Duckstein 1982. *Multiobjective decision analysis with engineering and business applications*. New York: Wiley.

Meredith, D. D., K. W. Wong, R. W. Woodhead & R. H. Wortman 1985. *Design and planning of engineering systems*, 2nd edn. Englewood Cliffs, NJ: Prentice-Hall.

Ossenbruggen, P. J. 1984. *Systems analysis for civil engineers*. New York: Wiley.

Smith, A. A., E. Hinton & R. W. Lewis 1983. *Civil engineering systems. Analysis and design*. New York: Wiley.

Stark, R. M. & R. L. Nicholls 1972. *Mathematical foundations for design. Civil engineering systems*. New York: McGraw-Hill.

Taha, H. A. 1982. *Operations research. An Introduction*, 3rd edn. New York: Macmillan.
Wagner, H. M. 1975. *Principles of operations research with applications to managerial decisions*, 2nd edn. Englewood Cliffs, NJ: Prentice-Hall.

Problems

7.1 Solve the following LP problems graphically:

(a) Max $Z = 3x_1 + 5x_2$
subject to
$$\begin{aligned} 2x_1 + x_2 &\leqslant 12 \\ 4x_1 + 5x_2 &\leqslant 40 \\ x_1, x_2 &\geqslant 0 \end{aligned}$$

(b) Min $Z = 10x_1 + 8x_2$
subject to
$$\begin{aligned} x_1 &\geqslant 4 \\ x_1 + 3x_2 &\geqslant 12 \\ x_1 + x_2 &\geqslant 8 \\ x_1, x_2 &\geqslant 0 \end{aligned}$$

(c) Max $Z = 4x_1 - 3x_2$
subject to
$$\begin{aligned} x_1 &\leqslant 10 \\ 2x_1 + 3x_2 &\geqslant 12 \end{aligned}$$
x_1, x_2 unrestricted in sign.

7.2 Formulate the following LP problems into canonical and standard form:

(a) Min $Z = 4x_1 + 2x_2$
subject to
$$\begin{aligned} 2x_1 + 3x_2 &\geqslant 9 \\ 3x_1 + x_2 &\geqslant 6 \\ x_1, x_2 &\geqslant 0 \end{aligned}$$

(b) Max $Z = x_1 + 2x_2 + 4x_3$
subject to
$$\begin{aligned} x_1 + x_3 &\leqslant 5 \\ 2x_1 - x_2 &\geqslant 4 \end{aligned}$$
$x_1, x_3 \geqslant 0$ x_2 unrestricted in sign

7.3 Solve Problems 7.1(a)–(c) using the simplex method.

7.4 Solve Problems 7.2(a) and (b) using the simplex method.

7.5 Portion of a road network is shown in Figure 7.12. The travel times and capacities in both directions for all roads are given in Table 7.8. During peak conditions, 5000 vehicles per hour are travelling from node 1 to node 6 and 4000 vehicles per hour are travelling from node 4 to node 2.

(a) Formulate an LP problem which could be used to find the volume of traffic in each direction on each road during peak conditions.
(b) Solve the LP problem using a standard computer package.

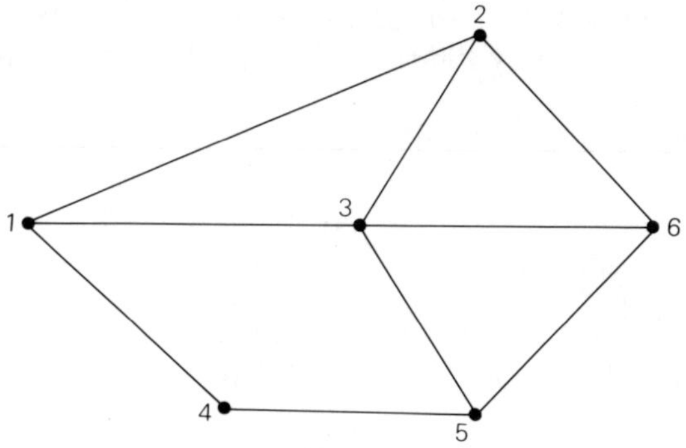

Figure 7.12 Portion of a road network.

Table 7.8 Travel times and capacities for a road network.

Road	Travel time (minutes)	Capacity (vehicles/h)
1–2	15	3000
1–3	10	4000
1–4	8	1000
2–6	6	3000
3–2	7	2000
3–5	6	3000
3–6	5	2000
4–5	9	3000
5–6	10	4000

7.6 A reservoir and pipe network is shown in Figure 7.13. The hydraulic grade line at locations C and D must be at or above RL 20 m and 0 m respectively. The elevation of the hydraulic grade line at B is to be determined. Three sizes of pipe are available according to Table 7.9.

Table 7.9 Pipe sizes and costs.

Pipe diameter (mm)	Cost ($/m length)
600	180
750	320
900	500

Each of the sections AB, BC, and BD may be made up of lengths of any or all pipe sizes. However, only 1000 m of 750 mm diameter pipe is available in time for the job.
The head loss in any length of pipe is given by:

$$h_\ell = 0.001 \frac{V^2 L}{D}$$

where h_ℓ is the head loss (m), V is the average velocity in pipe (m/s), L is the pipe length (m), and D is the pipe diameter (m).

Ignore the velocity head and head losses at entrance, exit, contraction, expansion, and bends.

(a) Set up an LP model which could be used to determine the lengths of pipe of each diameter to be used in each of the sections AB, BC, and BD so as to minimize the total cost.
(b) Solve the model using a suitable computer package.

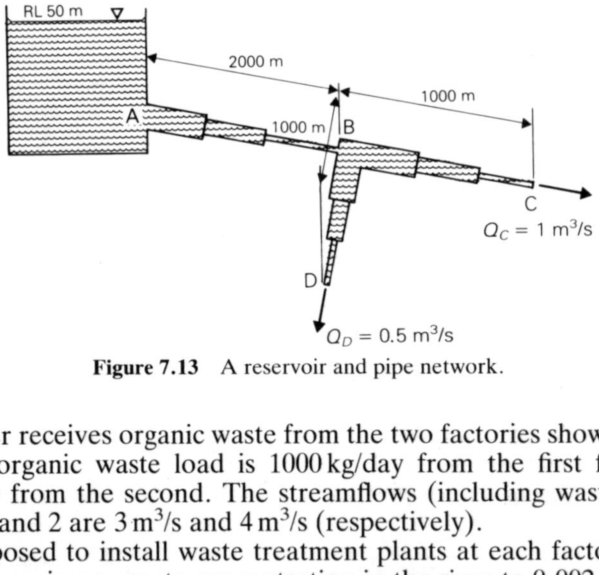

Figure 7.13 A reservoir and pipe network.

7.7 A river receives organic waste from the two factories shown in Figure 7.14. The organic waste load is 1000 kg/day from the first factory and 1400 kg/day from the second. The streamflows (including waste input) at locations 1 and 2 are 3 m^3/s and 4 m^3/s (respectively).

It is proposed to install waste treatment plants at each factory so as to reduce the maximum waste concentration in the river to 0.002 kg/m^3 at all locations. The cost of waste removal is \$0.30/kg at factory 1 and \$0.20/kg at factory 2. A maximum of 90% of the waste from either factory can be removed. Waste decomposition in the river may be ignored.

(a) Formulate an LP model which could be used to find the minimum cost combination of waste treatment plants at factories 1 and 2 such that the maximum waste concentration is not exceeded.
(b) Solve the model graphically.

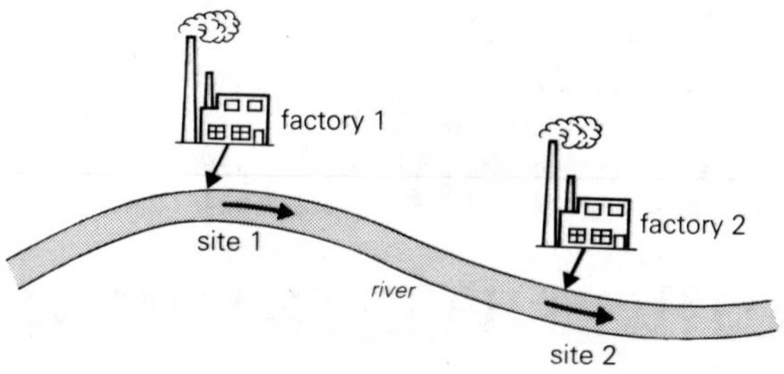

Figure 7.14 A river receiving organic waste from two factories.

7.8 Write the duals to Problems 7.1(a)–(c) and 7.2(a)–(b).

7.9(a) Write the dual to the LP model formulated in Problem 7.7.

(b) From the graphical solution of the primal, determine the optimum values of the dual variables. (Hint: the dual variables will be zero for those primal constraints which have some slack or surplus at the optimum solution. Why?)

(c) Verify the optimum values of the dual variables found in part (b) by solving the dual problem using the simplex method.

(d) Use the value of the dual variables to estimate the reduction in treatment cost if the maximum allowable waste concentration at site 2 is increased to $0.0022\,kg/m^3$.

8 Non-linear optimization and dynamic programming

In this chapter techniques for the solution of non-linear optimization problems are discussed. Many engineering planning and design problems contain a non-linear objective function or constraints. The specific technique of dynamic programming is useful for optimizing multistage decision problems.

8.1 Introduction

In Chapter 7 a special case of optimization called linear programming (LP) was examined. This applies to problems with a linear objective function and a set of linear constraints. Many general concepts of optimization can be demonstrated using LP. The assumption of linearity is often not appropriate to real engineering systems. In general, the objective function and/or constraints may be non-linear, if indeed they can be expressed in mathematical form.

There are many techniques available for the solution of non-linear optimization problems. In this chapter the techniques of differential calculus, Lagrange multipliers, separable programming, and dynamic programming are considered.

Many practical engineering problems do not lend themselves to the use of formal optimization techniques. In some cases it is possible to develop a mathematical model of the system but not to optimize it. Computer simulation (see Section 3.6) can then be used in a trial-and-error search for the optimum solution.

In many other design situations, formal optimization is not possible but an approximation to the optimum design is found by evaluating a range of alternative solutions.

8.2 Non-linear optimization

The general single-objective, non-linear optimization problem may be written as:

$$\text{Max} \quad Z = f(x_1, x_2, \ldots, x_n)$$

subject to

$$h_i(x_1, \ldots, x_n) \begin{Bmatrix} \leqslant \\ = \\ \text{or } \geqslant \end{Bmatrix} b_i \; (i=1, \ldots, m)$$

This problem differs from the general LP problem (Conditions 7.3–7.5) in three major respects:

- The feasible solution space for a non-linear problem may be either convex or non-convex. (This follows from the fact that the boundaries of the feasible region may, in general, be generated by non-linear functions. Figure 7.5 illustrates some non-convex feasible regions.)
- An optimum solution to a non-linear problem may occur at any point in the feasible region (not necessarily at an extreme point).
- It is possible for more than one 'optimum' to exist. The solution corresponding to the absolute maximum (or minimum) of the objective function is called the **global optimum**; all other optima are **local optima**.

Two examples are used to illustrate the second point.

Example 8.1 Consider the following problem:

$$\text{Min} \quad Z = (x_1 - 2)^2 + (x_2 - 3)^2$$

subject to:

$$2x_1 + x_2 \leqslant 14$$

$$x_1 + 3x_2 \leqslant 15$$

$$x_1, x_2 \geqslant 0$$

The feasible solution space and contours of the objective function for this problem are shown in Figure 8.1. The minimum value of Z equals zero and occurs at $x_1 = 2$, $x_2 = 3$. This is in the interior of the feasible region.

Example 8.2 Consider the following problem:

$$\text{Max} \quad Z = x_1 x_2$$

subject to:

$$x_2 + 0.1x_1^2 \leqslant 10$$

$$x_1 \leqslant 7$$

$$x_1, x_2 \geqslant 0$$

The feasible region and contours of Z are shown in Figure 8.2. In this

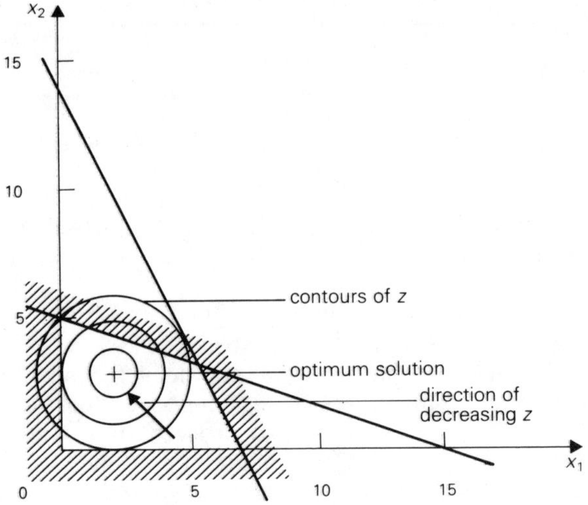

Figure 8.1 Optimum solution at interior of the feasible region.

case the optimum solution ($Z = 38.49$ at $x_1 = 5.77$, $x_2 = 6.67$) lies on the boundary of the feasible region but not at an extreme point.

In relation to the third point above, it is useful to think of the objective function surface as a range of mountains. In a maximization problem every peak is a local optimum, and the highest peak is the global optimum. Of course, in many problems, the objective function has only one peak and there are no local optima. Depending on the constraints this may or may not correspond to the optimum solution.

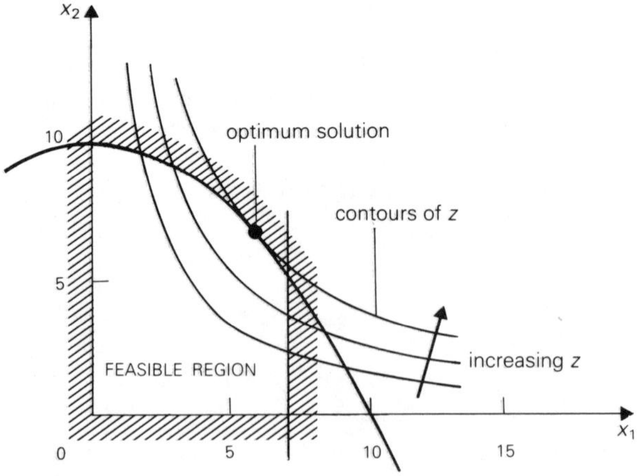

Figure 8.2 Optimum solution at boundary of the feasible region.

In linear programming, the feasible solution space is always convex, the optimum solution always occurs at an extreme point and local optima do not occur.

These differences make non-linear optimization considerably more difficult than linear optimization. In fact, there is no single solution technique that works for all non-linear optimization problems. Many specific techniques have been developed, each of which is applicable to a certain category of non-linear problems.

The techniques may be grouped into the following categories:

- the graphical technique
- classical optimization techniques
- numerical techniques

The graphical technique may be used for problems with only one or two decision variables. It involves plotting the constraints and the objective function in terms of the decision variables in order to identify the optimum solution. The technique is illustrated by Examples 8.1 and 8.2 considered earlier. It is generally only suitable for simple functions.

Classical optimization techniques utilize differential calculus. They may be used for unconstrained non-linear optimization problems if the non-linear functions are well behaved. They may also be used for simple constrained optimization problems (particularly those with equality constraints).

Numerical optimization techniques are often preferred to calculus-based techniques because of the difficulty in applying the latter to problems of reasonable size. A large number of numerical techniques are available for solving non-linear optimization problems.

The techniques considered in this chapter are:

- **classical techniques:** differential calculus, Lagrange multipliers
- **numerical techniques:** separable programming, dynamic programming

The interested reader should refer to Beightler *et al.* (1979) or Wagner (1975) for a more detailed treatment of various non-linear optimization techniques.

8.3 Unconstrained problems using calculus

The general unconstrained optimization problem is:

$$\text{Max } Z = f(x_1, x_2, \ldots, x_n) \tag{8.1}$$

where $f(x_1, x_2, \ldots, x_n)$ is any non-linear function of the n decision variables $x_1, \ldots, x_n$.

This is perfectly general, as a minimization problem can be converted into a maximization problem by multiplying the objective function by -1.

Firstly consider the single-variable case:

$$\text{Max } Z = f(x) \tag{8.2}$$

Classical differential calculus can be used to find the maxima of $f(x)$ provided that $f(x)$ and its first- and second-order derivatives exist and are continuous for all values of x.

A **necessary condition** for a particular solution x^o to be a maximum of $f(x)$ is:

$$\left.\frac{df(x)}{dx}\right|_{x = x^o} = 0 \tag{8.3}$$

The solutions to Equation (8.3) are called **stationary points** and may include maxima, minima, and points of inflection. A **point of inflection** is a point where $f(x)$ has zero slope but is neither a maximum nor a minimum. (For example, the function $f(x) = x^3$ has a point of inflection at $x = 0$ as shown in Figure 8.3.)

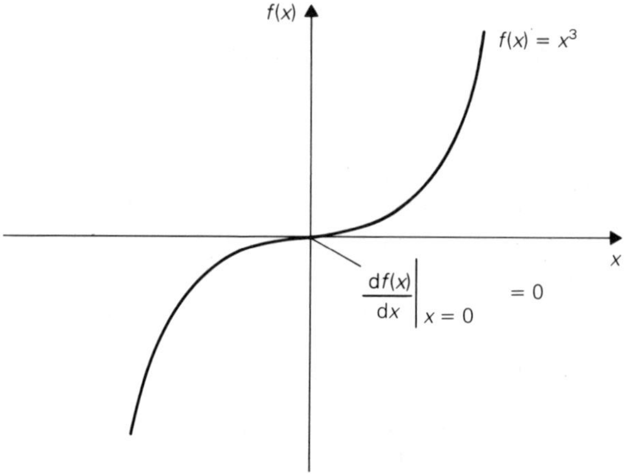

Figure 8.3 Point of inflection.

A **sufficient** condition for a stationary point x^o to be a maximum of $f(x)$ is:

$$\left.\frac{d^2f(x)}{dx^2}\right|_{x = x^o} < 0 \tag{8.4}$$

A sufficient condition for a minimum at x^o is:

$$\left.\frac{d^2f(x)}{dx^2}\right|_{x = x^o} > 0 \tag{8.5}$$

These conditions can be extended to the unconstrained multivariate case defined by Condition 8.1. For convenience define $\underline{\mathbf{X}} = (x_1, x_2, \ldots, x_n)$, then Condition 8.1 becomes:

$$\text{Max } Z = f(\underline{\mathbf{X}}) \tag{8.6}$$

As in the single-variable problem, differential calculus can be used to solve this problem provided $f(\underline{\mathbf{X}})$ and its first- and second-order partial derivatives exist and are continuous for all values of $\underline{\mathbf{X}}$.

A **necessary** condition for a particular solution $\underline{\mathbf{X}}^{\circ}$, to be a maximum of $f(\underline{\mathbf{X}})$ is:

$$\left.\frac{\partial f(\underline{\mathbf{X}})}{\partial x_j}\right|_{\underline{\mathbf{X}} = \underline{\mathbf{X}}^{\circ}} = 0 \qquad (j=1, \ldots, n) \tag{8.7}$$

The solutions to Equation (8.7) are called stationary points and include maxima, minima, points of inflection, and saddlepoints.

In order to determine which of the stationary points are maxima (or minima) it is necessary to examine the second-order partial derivatives of $f(\underline{\mathbf{X}})$. These are contained in the Hessian matrix, $\underline{\mathbf{H}}$, which is defined as follows:

$$\underline{\mathbf{H}} = \begin{bmatrix} \dfrac{\partial^2 f}{\partial x_1^2} & \dfrac{\partial^2 f}{\partial x_1 \partial x_2} & \cdots & \dfrac{\partial^2 f}{\partial x_1 \partial x_n} \\ \dfrac{\partial^2 f}{\partial x_2 \partial x_1} & \dfrac{\partial^2 f}{\partial x_2^2} & \cdots & \dfrac{\partial^2 f}{\partial x_2 \partial x_n} \\ \vdots & & & \vdots \\ \dfrac{\partial^2 f}{\partial x_n \partial x_1} & \dfrac{\partial^2 f}{\partial x_n \partial x_2} & \cdots & \dfrac{\partial^2 f}{\partial x^2_n} \end{bmatrix}$$

A **sufficient** condition for a stationary point, $\underline{\mathbf{X}}^{\circ}$, to be a minimum is that the Hessian matrix evaluated at $\underline{\mathbf{X}}^{\circ}$ is **positive definite**.

For $\underline{\mathbf{X}}^{\circ}$ to be a maximum, the Hessian matrix evaluated at $\underline{\mathbf{X}}^{\circ}$ should be **negative definite**.

A matrix is positive definite if all of its **principal minor determinants** λ_k $(k=1, \ldots, n)$ are positive.

The kth principal minor determinant of $\underline{\mathbf{H}}$ is defined as:

$$\lambda_k = \begin{vmatrix} h_{11} & h_{12} & \cdots & h_{1k} \\ h_{21} & h_{22} & \cdots & h_{2k} \\ \vdots & & & \\ h_{k1} & h_{k2} & \cdots & h_{kk} \end{vmatrix}$$

A matrix is negative definite if its kth principal minor determinant has the sign $(-1)^k$ for $k = 1, 2, \ldots, n$.

Example 8.3 A tank of volume 200 m^3 is to be made from stainless steel. It will have a cylindrical body and hemispherical ends as shown in Figure 8.4. If the material for the ends costs twice as much per square metre as that for the body, find the dimensions of the tank which minimize its cost.

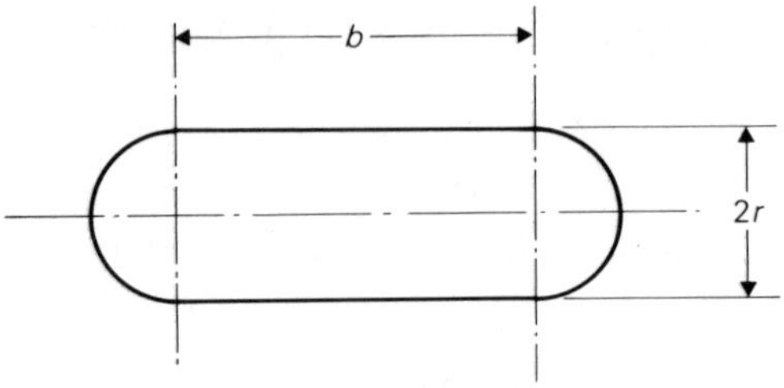

Figure 8.4 A stainless steel tank.

Solution Let the length of the body of the tank be b m and its radius be r m. If the cost per unit area of the body is C \$/m^2, then the total cost of the tank is:

$$Y = 2C\pi rb + 2C(4\pi r^2)$$

$$Y = 2C\pi rb + 8\,C\pi r^2 \tag{8.8}$$

As C is constant, it can be ignored in the optimization and the problem becomes:

$$\text{Min } Z = 2\pi rb + 8\pi r^2 \tag{8.9}$$

subject to the volume constraint:

$$\pi r^2 b + 4\pi r^3/3 = 200 \tag{8.10}$$

This constraint can be written as an equality because the minimum cost tank will not have excess volume.

Using Equation (8.10) we can substitute for b in the objective function

$$b = \frac{[200 - 4\pi r^3/3]}{\pi r^2} \tag{8.11}$$

Thus (8.9) becomes:

$$\text{Min} \quad Z = \frac{400}{r} + \frac{16}{3}\pi r^2 \tag{8.12}$$

which is an unconstrained single-variable optimization.

The necessary condition (8.3), is found by differentiating Z with respect to r,

$$\frac{dZ}{dr} = \frac{-400}{r^2} + \frac{32\pi r}{3} = 0$$

and then solving for r gives

$$r = 2.285\,\text{m}$$

and substituting into Equation (8.11),

$$b = 9.146\,\text{m}$$

Now checking the sufficiency condition (8.5):

$$\frac{d^2Z}{dr^2} = \frac{+800}{r^3} + \frac{32\pi}{3}$$

which is positive for all positive values of r, therefore the above solution is a minimum.

In practice the chosen dimensions would be rounded off, e.g. $b = 9\,\text{m}$, $r = 2.3\,\text{m}$.

8.4 Lagrange multipliers

For constrained non-linear optimization problems the conditions defined by Equations (8.7) do not necessarily hold at an optimum solution. For example consider the following single-variable problem:

$$\text{Max } Z = -x^2 + 8x - 6 \qquad (8.13)$$

$$\text{subject to} \quad x \leqslant 3 \qquad (8.14)$$

As shown in Figure 8.5, the objective function has a maximum value of 10 at $x = 4$. The derivative of Z is given by

$$\frac{dz}{dx} = -2x + 8 \qquad (8.15)$$

which is zero at $x = 4$.

However, the maximum value of Z in the constrained problem occurs when $x = 3$. Substitution into Equation (8.15) verifies that the derivative is not zero at this point (see Figure 8.5).

Thus the necessary and sufficient conditions for an unconstrained problem are not applicable in the constrained case.

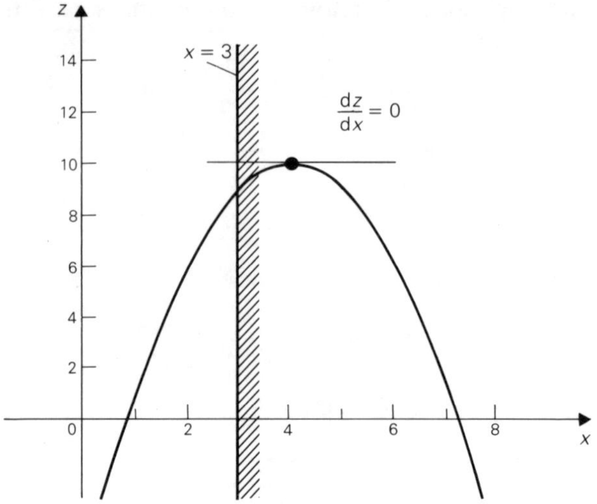

Figure 8.5 Constrained optimization.

Non-linear optimization problems with non-linear *equality* constraints can be solved by the following techniques:

- the method of substitution
- the Lagrange multiplier technique

The method of substitution was used in Example 8.3. It can be used if the constraints have a relatively simple form which allows variables to be eliminated from the objective function by substitution using the constraint equations. The problem is then converted into an unconstrained optimization problem.

The Lagrange multiplier technique is a more general method. It requires the non-linear functions in the objective function and constraints to be continuous and have continuous first- and second-order partial derivatives with respect to all variables.

Consider the following problem:

$$\text{Max} \quad Z = f(\underline{\mathbf{X}}) \tag{8.16}$$

subject to

$$g_i(\underline{\mathbf{X}}) = b_i \qquad (i=1, \ldots, m) \tag{8.17}$$

The stationary points to this problem can be identified by forming the Lagrangian function:

$$L(\underline{\mathbf{X}},\underline{\boldsymbol{\lambda}}) = f(\underline{\mathbf{X}}) - \sum_{i=1}^{m} \lambda_i \{g_i(\underline{\mathbf{X}}) - b_i\} \tag{8.18}$$

The new variables λ_i $(i=1, \ldots, m)$ are called the **Lagrange multipliers**. Each one corresponds to a constraint in the original problem.

The **necessary** conditions for a maximum may be found by setting the partial derivatives of the Lagrangian function with respect to x_j (j=1, . . ., n) and λ_i (i=1, . . ., m) equal to zero, i.e.

$$\frac{\partial L}{\partial x_j} = 0 \qquad (j=1, \ldots, n) \tag{8.19}$$

$$\frac{\partial L}{\partial \lambda_i} = 0 \qquad (i=1, \ldots, m) \tag{8.20}$$

Conditions (8.19) and (8.20) give $(m+n)$ non-linear equations in $(m+n)$ unknowns. In many cases they can be solved for x_j (j=1, . . ., n) and λ_i (i=1, . . ., m).

Sufficient conditions also exist for Equations (8.19) and (8.20) to define a maximum. Further information, including a proof of the Lagrange multiplier technique, is given in Taha (1982).

Example 8.4 Rework Example 8.3 using the Lagrange multiplier technique. Assume that C = \$100/m^2.

Solution Rewriting Equations (8.8) and (8.10),

$$\text{Min} \quad Y = 200\pi rb + 800\pi r^2 \tag{8.21}$$

subject to

$$\pi r^2 b + 4\pi r^3/3 = 200 \tag{8.22}$$

In this case there is only one constraint, so the Lagrangian becomes

$$L(\underline{\mathbf{X}},\underline{\boldsymbol{\lambda}}) = 200\pi rb + 800\pi r^2 - \lambda\{\pi r^2 b + 4\pi r^3/3 - 200\} \tag{8.23}$$

Differentiating with respect to r, b, and λ gives

$$\frac{\partial \lambda}{\partial r} = 200\pi b + 1600\pi r - 2\lambda\pi rb - 4\pi r^2 = 0 \tag{8.24}$$

$$\frac{\partial L}{\partial b} = 200\pi r - \lambda\pi r^2 = 0 \tag{8.25}$$

$$\frac{\partial L}{\partial \lambda} = \pi r^2 b + 4\pi r^3/3 - 200 = 0 \tag{8.26}$$

Solving Equations (8.24)–(8.26) gives

$$r = 2.285\,\text{m}, \; b = 9.146\,\text{m}$$

as before, with λ = \$87.5/m^3 and Y = \$26 253.

It should be noted that the equations formed by differentiating the Lagrangian with respect to the Lagrange multipliers are always the original constraints to the problem (compare Equations (8.22) and (8.26)).

The values of the Lagrange multipliers at optimality have particular significance. In fact it can be shown (Taha 1982) that, at optimality,

$$\lambda_i = \frac{\partial Z}{\partial b_i}$$

That is, the optimum value of the Lagrange multiplier corresponding to the ith constraint equals the change in objective function per unit change in the right-hand side of the ith constraint.

In the above example the optimum value of λ is \$87.5/m^3. This indicates that a small change, ε, in the right-hand side of Equation (8.22) (i.e. the volume of the tank) will cause the cost of the tank to increase by \$87.5(ε).

By referring to Section 7.11 it can be seen that Lagrange multipliers are in fact the **dual variables** of the optimization problem.

The Lagrange multiplier technique can be extended to non-linear optimization problems with inequality constraints but this is beyond the scope of the present text. The interested reader is referred to Taha (1982) or Smith *et al.* (1983).

8.5 Separable programming

Separable programming is an extension of linear programming. It works by converting a non-linear optimization problem into an LP problem which approximates the original non-linear one. Separable programming can only be applied to separable functions.

A function $f(x_1, x_2, \ldots, x_n)$ is said to be **separable** if it can be expressed as the sum of n single-variable functions $f_1(x_1), \ldots, f_j(x_j), \ldots, f_n(x_n)$.

Although separable programming can be used to approximate any continuous non-linear function it is most effective when the following special conditions apply:

- The individual functions, $f_j(x_j)$, of the objective function are concave for a maximization problem or convex for minimization problems.
- The constraints define a convex feasible region.

A **convex function** is one which has the property that a straight line joining any two points on the function always lies on or above the function. For a **concave function** a straight line joining any two points on the function always lies on or below the function. Examples of convex and concave functions are shown in Figure 8.6.

A **convex region** has been defined in Section 7.4.

If the properties defined above do not apply, separable programming involves the use of integer variables. This is beyond the scope of this book. The interested reader is referred to Taha (1982).

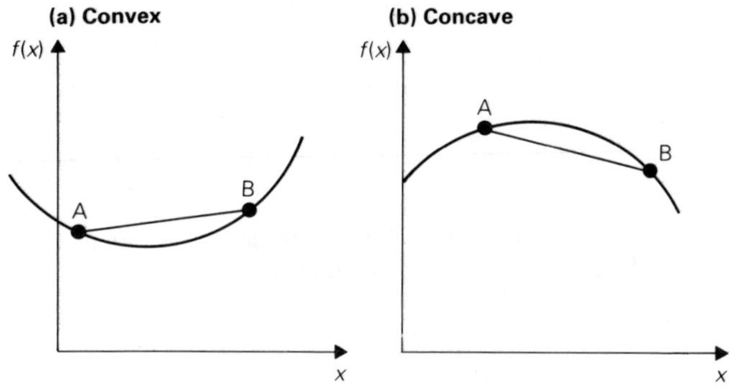

Figure 8.6 Convex and concave functions.

Separable programming always yields a **global** optimum to the approximate linear problem.

To illustrate the separable programming technique, consider the minimization of the convex objective function shown in Figure 8.7. The non-linear function is approximated by a series of straight line segments between selected values of x, namely $\lambda_1, \lambda_2, \ldots, \lambda_K$ which must be selected to span the full range of likely values of x.

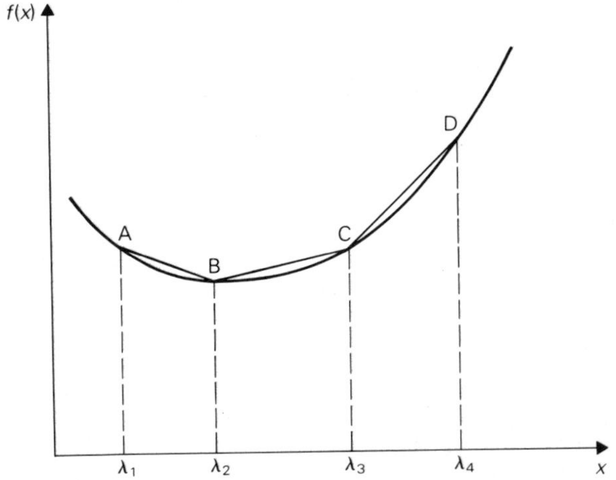

Figure 8.7 A convex objective function.

We define a new set of variables $w_1, \ldots, w_K$ such that

$$x = \sum_{k=1}^{K} w_k \lambda_k \tag{8.27}$$

$$\sum_{k=1}^{K} w_k = 1 \tag{8.28}$$

and

$$w_k \geqslant 0 \qquad (k = 1, \ldots, K) \tag{8.29}$$

In the final solution x is a weighted average of the values $(\lambda_1, \ldots, \lambda_K)$ where $w_k (k = 1, \ldots, K)$ are the relative weights. Provided at most two adjacent values of w_k are non-zero, $f(x)$ can be approximated by

$$f(x) = \sum_{k=1}^{K} w_k f(\lambda_k) \tag{8.30}$$

For example, if $w_2 = w_3 = 0.5$ and all other values of w_k are zero, Equations (8.27) and (8.30) give:

$$x = 0.5\lambda_2 + 0.5\lambda_3$$

$$f(x) = 0.5f(\lambda_2) + 0.5f(\lambda_3)$$

which is a point on the straight line midway between points A and B in Figure 8.7.

It can be shown that, at most, two adjacent weights w_k will be non-zero when using LP and the special conditions above apply.

This can be demonstrated intuitively for the case of minimizing a convex function as shown in Figure 8.8. Suppose as a particular solution we obtain $w_1 = 0.6$ and $w_3 = 0.4$ with all other $w_k = 0$. Then

$$x = 0.6\,\lambda_1 + 0.4\,\lambda_3$$

and

$$f(x) = 0.6f(\lambda_1) + 0.4f(\lambda_3)$$

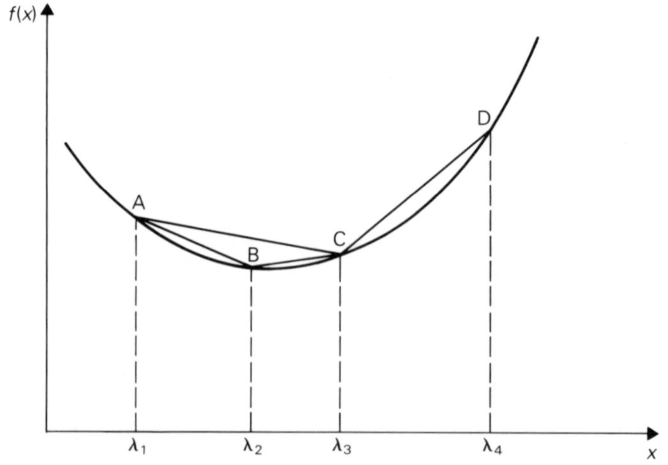

Figure 8.8 A convex objective function.

This is a point on the straight line segment between points A and C. Clearly the line segments AB and BC lie below this and an LP solution which is minimizing $f(x)$ will achieve a solution on one of the lower line segments – i.e. with only two adjacent weights non-zero.

Example 8.5 A cylindrical tank is to be made from a 6 m square sheet of metal as shown in Figure 8.9. The tank will be h m high and d m in diameter. The height of the tank must not be more than twice its diameter. Use separable programming to find the dimensions of the tank which maximize its volume.

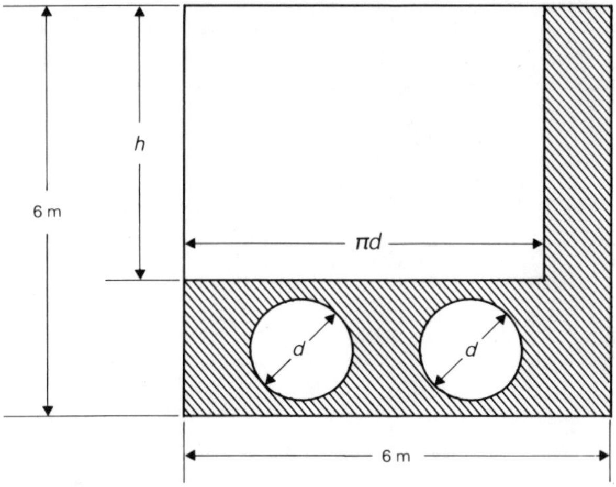

Figure 8.9 Plan for side and ends of a cylindrical tank.

Solution The optimization problem is:

$$\text{Max } V = \pi d^2 h/4 \qquad (8.31)$$

subject to

$$\pi d \leqslant 6$$

i.e.

$$d \leqslant 1.910 \qquad (8.32)$$

$$h + d \leqslant 6 \qquad (8.33)$$

$$h/d \leqslant 2$$

i.e.

$$h - 2d \leqslant 0 \qquad (8.34)$$

$$h, d \geqslant 0 \qquad (8.35)$$

The objective function can be made separable by taking its logarithm (to the base 10), i.e.

$$\log V = \log(\pi/4) + 2\log d + \log h \tag{8.36}$$

Clearly, maximizing log V is equivalent to maximizing V. As log d and log h are concave functions in a maximization objective, we can use separable programming without integer variables.

It is now necessary to approximate log d and log h over a suitable range. First consider log h. From Equation (8.33) it is obvious that h must be less than 6 m. An approximation to log h for h between 0.5 m and 6 m is shown in Figure 8.10. Values for λ_k and $\log_{10}\lambda_k$ are given in Table 8.1. Applying Equations (8.27)–(8.30) to approximate the function log (h) gives

$$h = 0.5w_1 + w_2 + 2w_3 + 4w_4 + 6w_5 \tag{8.37}$$

$$w_1 + w_2 + w_3 + w_4 + w_5 = 1 \tag{8.38}$$

$$w_i \geqslant 0 \qquad (i = 1, \ldots, 5) \tag{8.39}$$

$$\log h = -0.301w_1 + 0.w_2 + 0.301w_3 + 0.602w_4 + 0.778w_5 \tag{8.40}$$

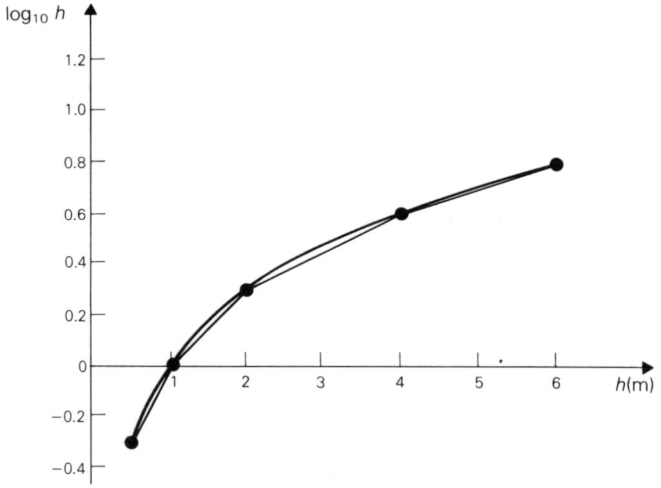

Figure 8.10 Approximation to $\log_{10}h$.

As d must be less than or equal to 1.91 m, log d will be approximated over the range 0.5 m–2 m. Using Table 8.1, this gives

$$d = 0.5u_1 + u_2 + 2u_3 \tag{8.41}$$

$$u_1 + u_2 + u_3 = 1 \tag{8.42}$$

$$u_j \geqslant 0 \qquad (j = 1, \ldots, 3) \tag{8.43}$$

$$\log d = -0.301u_1 + 0.u_2 + 0.301u_3 \tag{8.44}$$

Thus, combining Equations (8.36), (8.40), and (8.44), the objective function becomes

$$\text{Max} \log V = \log (\pi/4) - 0.602u_1 + 0.602u_3$$

$$- 0.301w_1 + 0.301w_3 + 0.602w_4 + 0.778w_5$$

Subject to the constraints (8.32), (8.33), (8.34), (8.35), (8.37), (8.38), (8.39), (8.41), (8.42) and (8.43).

Solving this problem using LP gives the solution: $h = 3.82$ m, $d = 1.91$ m, and $V = 10.95$ m^3 which can be shown to be the global optimum.

Table 8.1 Values of λ_k and $\log_{10}\lambda_k$ for the cylindrical tank problem.

k	λ_k	$\log_{10}\lambda_k$
1	0.5	−0.301
2	1.0	0.000
3	2.0	0.301
4	4.0	0.602
5	6.0	0.778

8.6 Dynamic programming

Dynamic programming is a non-linear optimization technique which is applicable to a wide class of multi-stage decision problems. The term **dynamic** indicates the application of the technique to problems which involve movement from one point in time or space to another point by means of a number of intermediate stages (intervals of time or distance). At each stage a decision has to be taken, i.e. a choice from several alternatives has to be made, and the problem is to find the optimum set of decisions.

The method will be illustrated by the following example. Goods are to be transported by truck from city A to city K via the highway network shown schematically in Figure 8.11. The cost of transport on each length of highway in

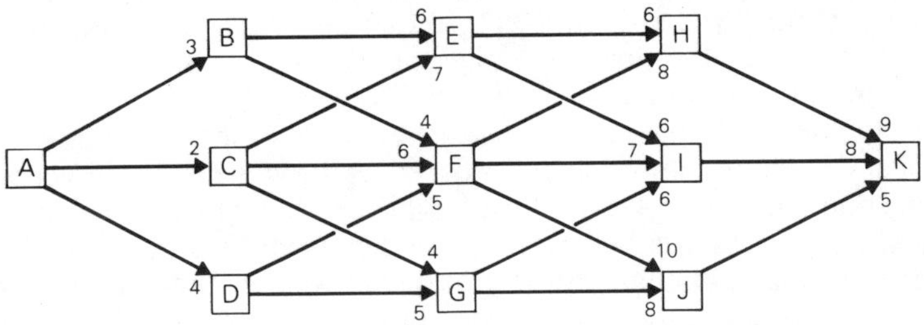

Figure 8.11 Highway routes and travel costs in dollars.

dollars is indicated on the figure. The problem is to find the minimum-cost route from city A to city K.

Inspection of Figure 8.11 indicates that, regardless of the route taken, the truck must pass through three other cities. Thus we may consider the problem to consist of four sequential decisions. We say that the problem consists of four stages. A **stage** is defined as the interval which separates consecutive decision points.

One method for determining the minimum-cost route would be to evaluate all alternatives. This is called **complete enumeration**, and in this case it would involve the costing of 17 feasible routes. For a larger problem containing, for example, 6 stages and 10 cities at each stage, there would be 10^5 alternative paths to evaluate. Clearly a more efficient approach is needed. Dynamic programming provides such an approach.

Dynamic programming is based on the **Principle of Optimality**, which may be stated as follows:

An optimal policy must have the property that regardless of the decisions leading to a particular state, the remaining decisions must constitute an optimal sequence for leaving that state.

As discussed in Chapter 3, the 'state' of a system summarizes concisely its entire previous history and therefore allows the future behaviour to be determined for any future input. In this example the states are the cities passed through.

When applied to the current problem, the principle of optimality means that if the optimum route passes through state F, it must conclude with the optimum route from F to K.

Thus the principle of optimality can be applied one stage at a time through the network in order to identify the minimum-cost path. Although it is possible to proceed either backwards or forwards in time, it is usually advantageous to work backwards. The present problem will be worked backwards.

With one stage to go, the truck could be in state H, I, or J. There is no choice of destination, which must be K. Nevertheless the calculations are set up in tabular form in order to be consistent with the rest of the solution. The information with one stage to go is summarized in Table 8.2(a). Regardless of the initial state the decision must be to move to K. The optimum decisions and corresponding minimum costs for these trips are summarized in Table 8.2(a).

With two stages to go the truck could be in city E, F, or G and a decision must be made regarding the next state. The minimum cost for each decision is the sum of the cost for the first stage plus the minimum cost for the remaining stages. This information is summarized in Table 8.2(b).

If starting in city E there is no route to get to city J, so the choice is between cities H and I. The minimum cost of proceeding through H is the direct cost ($6) plus the minimum cost from H onwards which is obtained from Table 8.2(a) as $9. Similarly, the minimum cost of proceeding from E through I is the direct cost ($6) plus the minimum cost from I onwards ($8). The optimum decision from E corresponds to the smallest cost, namely $14 via city I. Similar calculations are made for the initial states of F and G.

Table 8.2 Decision tables for highway route problem.

(a) With one stage remaining

Initial state	Decision: *K*	Optimum decision	Minimum cost ($)
H	9	K	9
I	8	K	8
J	5	K	5

(b) With two stages remaining

Initial state	Decision: H	I	J	Optimum decision	Minimum cost ($)
E	6 + 9	6 + 8	—	I	14
F	8 + 9	7 + 8	10 + 5	I or J	15
G	—	6 + 8	8 + 5	J	13

(c) With three stages remaining

Initial state	Decision: E	F	G	Optimum decision	Minimum cost ($)
B	6 + 14	4 + 15	—	F	19
C	7 + 14	6 + 15	4 + 13	G	17
D	—	5 + 15	5 + 13	G	18

(d) With four stages remaining

Initial state	Decision: B	C	D	Optimum	Minimum
A	3 + 19	2 + 17	4 + 18	C	19

Table 8.2(b) indicates the optimum decision from all cities with two stages to go. Thus if a truck driver should ever find himself in city G, the optimum decision is to go to city J, and so on.

At this point we have evaluated all possible paths over the last two stages. Now with three stages to go, dynamic programming enables us to evaluate only a subset of all possible paths.

Table 8.2(c) summarizes the calculations with three stages to go. If starting in city B, the choice is between proceeding to city E or F. The minimum cost of travelling through E is the sum of the direct cost ($6) plus the minimum cost from E onwards. The latter term is given in Table 8.2(b) as $14. Similarly, the minimum cost of travelling through city F may be assessed. Thus the optimum route from B is via city F at a cost of $19. The minimum cost route from states C and D are also indicated in Table 8.2(c).

Finally, with four stages to go the initial state is A and the choice is between B, C, and D. The minimum cost via B is the direct cost of $3 plus the minimum cost from B onwards ($19 from Table 8.2(c)). The minimum-cost route is via C at a cost of $19.

The optimum route through the network can be determined by working forwards from the starting state A using Table 8.2. Table 8.2(d) indicates that from A we should proceed to C. Table 8.2(c) shows that G is the optimum choice from C. From G we go to J (Table 8.2(b)) and from J to K (Table 8.2(a)). Therefore the minimum cost route is A-C-G-J-K at a cost of $19. Note also that the dynamic programming solution identifies the optimum solution extending from any non-optimal intermediate state. For example, should an incorrect initial decision lead to B instead of C, the optimal subsequent sequence is B-F-I-K or B-F-J-K.

The benefits of using dynamic programming to solve this problem should be apparent from Table 8.2. With three stages to go only seven comparisons had to be made, even though there are seventeen possible routes through the last three stages. For larger problems, the advantage of using dynamic programming is more marked. For example, a problem with six stages and ten states at each stage contains 10^5 paths, but dynamic programming involves only 410 comparisons.

It should be noted that dynamic programming involves the implicit evaluation of all possible paths. Therefore it always achieves the **global optimum**. It is clearly a very powerful technique for problems which can be organized into a suitable form.

The dynamic recursive equation

Although computer packages for solving general dynamic programming problems are available, it is often better for the analyst to write a computer program for the specific problem being solved. The tabular form of solution given above is suitable for hand solution of dynamic programming problems. Computer codes for solving dynamic programming problems rely on the development of a suitable recursive equation as follows:

Let $c_n(i,j)$ be the cost of proceeding from state i with n stages to go to state j with $(n-1)$ stages to go.
Let $f_n(i)$ be the cost of the optimum policy to the end of the final stage when starting in state i with n stages to go.
Let N be the total number of stages in the problem.

Then the dynamic recursive relationship for a minimization problem may be written as:

$$f_n(i) = \min_j \left[c_n(i,j) + f_{n-1}(j) \right] \qquad n=1, \ldots, N \qquad (8.45)$$

Thus, the cost of the optimum policy when starting in state i with n stages to go is the minimum over all feasible states j of going from i to j *plus* the cost of the optimum policy when in state j with $n-1$ stages to go.

The use of the dynamic recursive equation will be demonstrated for a problem of reservoir operation.

Example 8.6 A single reservoir of capacity 6 units is situated on a river and is used to regulate the releases to downstream irrigators. A maximum of 4 units can be released in one season. An optimum release policy is required over the next four seasons beginning with winter. The economic benefits to irrigators for various releases are shown in Table 8.3. The expected inflows to the reservoir over the next four seasons are 4, 3, 1, and 2 units, respectively. At the beginning of winter the reservoir starts with 3 units in storage and must have at least 3 units in storage at the end of autumn. Dynamic programming should be used to find the optimum release policy.

Table 8.3 Seasonal benefits for release of water for irrigation.

Release (units)	Benefits ($M) for release during:			
	winter	spring	summer	autumn
1	0.2	1.5	1.3	0.5
2	0.4	2.5	2.3	1.0
3	0.5	3.3	3.0	1.2
4	0.6	3.8	3.3	1.3

Solution The stages in this problem are the four seasons; the state variable is the storage in the reservoir at the end of each season as this summarizes the effects of all previous inputs and decisions. The objective function (to be maximized) is the benefit of water released for irrigation.

Figure 8.12 shows the feasible states for this problem. Given the initial storage of 3 units and the winter inflow of 4 units the minimum storage at the end of winter is 3 units corresponding to a release of 4 units. The maximum storage is 6 units which corresponds to the release of 1 unit. (A release of zero units is not possible because a spill of 1 unit will occur in any case.)

At the end of spring a minimum storage of 2 units can be achieved by releasing 4 units from an initial storage of 3 units.

At the end of summer, the minimum feasible storage is 1 unit which can be achieved via several paths. Although a storage of zero units can be achieved it is not possible to come up to the desired storage of 3 units at the end of autumn from this state.

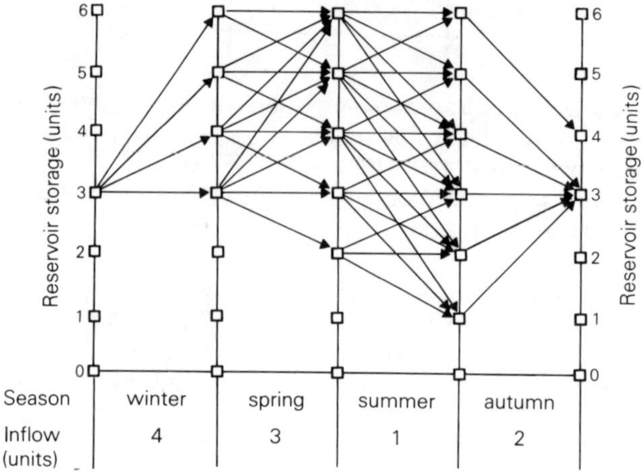

Figure 8.12 Feasible states for the reservoir release problem.

The storage at the end of autumn will be 3 units in most cases as there is no advantage in having more water in storage. However, if the storage at the end of summer is 6 units, the minimum achievable storage at the end of autumn is 4 units.

The dynamic recursive equation (8.45) will be applied backwards in time to this problem. In this case, let

i_n be the reservoir storage with n stages remaining,

x_n be the reservoir release during the current stage with n stages remaining, and

q_n the inflow during the current stage with n stages remaining (known). Clearly

$$i_{n-1} = i_n - x_n + q_n \tag{8.46}$$

i.e. the reservoir storage with $(n-1)$ stages remaining is the storage with n stages remaining minus the release plus the inflow during the current stage. Therefore, the dynamic recursive relation for this problem becomes:

$$f_n(i_n) = \max_{0 \leq x_n \leq 4} \left[c_n(x_n) + f_{n-1}(i_{n-1}) \right] \quad \begin{matrix}(n=0, 1, \ldots, 4) \\ \text{all feasible } i_n\end{matrix} \tag{8.47}$$

where $f_n(i_n)$ are the maximum benefits when in state i with n stages remaining and $c_n(x_n)$ is the benefit of releasing x_n when n stages remain.
Substituting Equation (8.46) into Equation (8.47),

$$f_n(i_n) = \max_{0 \leq x_n \leq 4} \left[c_n(x_n) + f_{n-1}(i_n - x_n + q_n) \right] \quad \begin{matrix}(n=0, 1, \ldots, 4) \\ \text{all feasible } i_n\end{matrix} \tag{8.48}$$

Now to apply Equation (8.48) backwards in time, we start by defining

$$f_o(i_o) = 0 \qquad (i_o=0, 1, \ldots, 6)$$

(a) *With one stage remaining:*

$$f_1(i_1) = \max_{0 \leq x_1 \leq 4} \left[c_1(x_1) + f_o(i_1 - x_1 + 2) \right] \quad \text{(for } i_1{=}1, \ldots, 6\text{)}$$

where values of $c_1(x_1)$ are given in the right-hand column of Table 8.3. Clearly, there is no choice at this stage, so we simply determine:

$f_1(1) = [c_1(0) + f_0(1{-}0{+}2)] = [c_1(0) + f_0(3)] = 0 + 0 = \$0$ (with $x_1^*(1) = 0$)

where $x_n^*(i_n)$ denotes the optimum release when in state i_n with n stages remaining.

$f_1(2) = [c_1(1) + f_0(2{-}1{+}2)] = 0.5 + 0 = \$0.5\text{M}$ (with $x_1^*(2) = 1$)

$f_1(3) = [c_1(2) + f_0(3{-}2{+}2)] = 1.0 + 0 = \$1.0\text{M}$ (with $x_1^*(3) = 2$)

$f_1(4) = [c_1(3) + f_0(4{-}3{+}2)] = 1.2 + 0 = \$1.2\text{M}$ (with $x_1^*(4) = 3$)

$f_1(5) = [c_1(4) + f_0(5{-}4{+}2)] = 1.3 + 0 = \$1.3\text{M}$ (with $x_1^*(5) = 4$)

$f_1(6) = [c_1(4) + f_0(6{-}4{+}2)] = 1.3 + 0 = \$1.3\text{M}$ (with $x_1^*(6) = 4$)

These calculations are summarized in Table 8.4(a).

(b) *With two stages remaining*, Equation (8.48) becomes:

$$f_2(i_2) = \max_{0 \leq x_2 \leq 4} \left[c_2(x_2) + f_1(i_2 - x_2 + 1) \right] \quad \text{(for } i_2{=}2, 3, \ldots, 6\text{)}$$

with $i_2 = 2$ only releases of 0, 1, or 2 units are feasible.
Comparing these,

$$f_2(2) = \max \begin{bmatrix} c_2(0) + f_1(2{-}0{+}1) \\ c_2(1) + f_1(2{-}1{+}1) \\ c_2(2) + f_1(2{-}2{+}1) \end{bmatrix} = \max \begin{bmatrix} 0 + 1.0 \\ 1.3 + 0.5 \\ 2.3 + 0 \end{bmatrix}$$

$$= \$2.3\text{M} \qquad \text{(with } x_2^*(2) = 2\text{)}$$

Note that the values of $f_1(1)$–$f_1(3)$ are given in part (a) above and in Table 8.4(a). The calculations with two stages remaining are summarized in Table 8.4(b).

(c) *With three stages remaining*, Equation (8.48) becomes

$$f_3(i_3) = \max_{0 \leq x_3 \leq 4} \left[c_3(x_3) + f_2(i_3 - x_3 + 3) \right] \quad \text{(for } i_3 = 3,4,5,6\text{)}$$

The calculations are summarized in Table 8.4(c).

Table 8.4 Decision tables for reservoir release problem.

(a) With one stage remaining

Initial storage i_1	Release (x_1) 0	1	2	3	4	Optimum release $x_1{}^*(i_1)$	Maximum benefit ($M) $f_1(i_1)$
1	0 + 0	—	—	—	—	0	0
2	—	0.5 + 0	—	—	—	1	0.5
3	—	—	1.0 + 0	—	—	2	1.0
4	—	—	—	1.2 + 0	—	3	1.2
5	—	—	—	—	1.3 + 0	4	1.3
6	—	—	—	—	1.3 + 0	4	1.3

(b) With two stages remaining

Initial storage i_2	Release (x_2) 0	1	2	3	4	Optimum release $x_2{}^*(i_2)$	Maximum benefit ($M) $f_2(i_2)$
2	0 + 1.0	1.3 + 0.5	2.3 + 0	—	—	2	2.3
3	0 + 1.2	1.3 + 1.0	2.3 + 0.5	3.0 + 0	—	3	3.0
4	0 + 1.3	1.3 + 1.2	2.3 + 1.0	3.0 + 0.5	3.3 + 0	3	3.5
5	0 + 1.3	1.3 + 1.3	2.3 + 1.2	3.0 + 1.0	3.3 + 0.5	3	4.0
6	0 + 1.3	1.3 + 1.3	2.3 + 1.3	3.0 + 1.2	3.3 + 1.0	4	4.3

(c) With three stages remaining

Initial storage i_3	Release (x_3) 0	1	2	3	4	Optimum release $x_3{}^*(i_3)$	Maximum benefit ($M) $f_3(i_3)$
3	0 + 4.3	1.5 + 4.0	2.5 + 3.5	3.3 + 3.0	3.8 + 2.3	3	6.3
4	0 + 4.3	1.5 + 4.3	2.5 + 4.0	3.3 + 3.5	3.8 + 3.0	3 or 4	6.8
5	0 + 4.3	1.5 + 4.3	2.5 + 4.3	3.3 + 4.0	3.8 + 3.5	3 or 4	7.3
6	0 + 4.3	1.5 + 4.3	2.5 + 4.3	3.3 + 4.3	3.8 + 4.0	4	7.8

(d) With four stages remaining

Initial storage i_4	Release (x_4) 0	1	2	3	4	Optimum release $x_4{}^*(i_4)$	Maximum benefit ($M) $f_4(i_4)$
3	0 + 7.8	0.2 + 7.8	0.4 + 7.3	0.5 + 6.8	0.6 + 6.3	1	8.0

(d) *With four stages remaining*, the initial storage is 3 units and Equation (8.48) becomes

$$f_4(3) = \max_{0 \leq x_3 \leq 4} \left[c_4(x_4) + f_3(3 - x_3 + 4) \right]$$

The calculations are summarized in Table 8.4(d).

To find the optimum release policy over the four seasons we work backwards through Table 8.4 as follows: release 1 unit in winter (leaving a final storage of 6), release 4 units in spring (final storage = 5), release 3 units in summer (final storage = 3), and release 2 units in autumn (final storage = 3). The economic benefits are \$8.0M for the year.

In actual reservoir operations the future inflows are usually not known with certainty, but only in a probabilistic sense. In this case an extension of dynamic programming called **stochastic dynamic programming** may be applied.

Dimensionality in dynamic programming

The examples considered above both contained only one state variable. Many real problems contain more than one state variable. An example is the operation of a system of n interconnected reservoirs. Here the state of the system at any time is summarized by the storage in each of the reservoirs and there are n state variables.

An increase in the number of state variables significantly increases the amount of computational effort to solve dynamic programming problems. This has been referred to as the **curse of dimensionality**. It may be illustrated by the reservoir operating problem. Suppose it is desired to apply dynamic programming to a single reservoir. If we divide the volume of the reservoir into 10 discrete levels, there are 10 feasible states at each stage. Suppose we now consider a system of n reservoirs, with the storage in each one being able to take on any one of 10 discrete values. The total number of states at each stage of this problem is 10^n. For example, if $n=4$ there are 10 000 feasible states at each stage. In practical terms, dynamic programming problems are limited to three or four state variables because of this difficulty with dimensionality.

References

Beightler, C., D. Phillips & D. Wilde 1979. *Foundations of optimization*, 2nd edn. Englewood Cliffs, NJ: Prentice-Hall.

deNeufville, R. & J. H. Stafford 1971. *Systems analysis for engineers and managers*. New York: McGraw-Hill.

Meredith, D. D., K. W. Wong, R. W. Woodhead & R. H. Wortman 1985. *Design and planning of engineering systems*. 2nd edn. Englewood Cliffs, NJ: Prentice-Hall.

Ossenbruggen, P. J. 1984. *Systems analysis for civil engineers*. New York: Wiley.

Smith, A. A., E. Hinton & R. W. Lewis 1983. *Civil engineering systems. Analysis and design*. New York: Wiley.

Stark, R. M. & R. L. Nicholls 1972. *Mathematical foundations for design. Civil engineering systems*. New York: McGraw-Hill.

Taha, H. A. 1982. *Operations research. An introduction*, 3rd edn. New York: Macmillan.
Wagner, H. M. 1975. *Principles of operations research with applications to managerial decisions*, 2nd edn. Englewood Cliffs, NJ: Prentice-Hall.

Problems

8.1 Determine the span lengths x_1, x_2, and x_3 (in metres) that minimize the cost of the horizontal bridge shown in Figure 8.13, given the following information on costs:

$$\begin{aligned} \text{cost of abutments} &= \$5000\text{/m of height} \\ \text{cost of piers} &= \$4000\text{/m of height} \\ \text{weight of steel for girders} &= 150x_i^2\,\text{kg}\ (i = 1, 2, 3) \\ \text{cost of steel} &= \$1.20\text{/kg} \end{aligned}$$

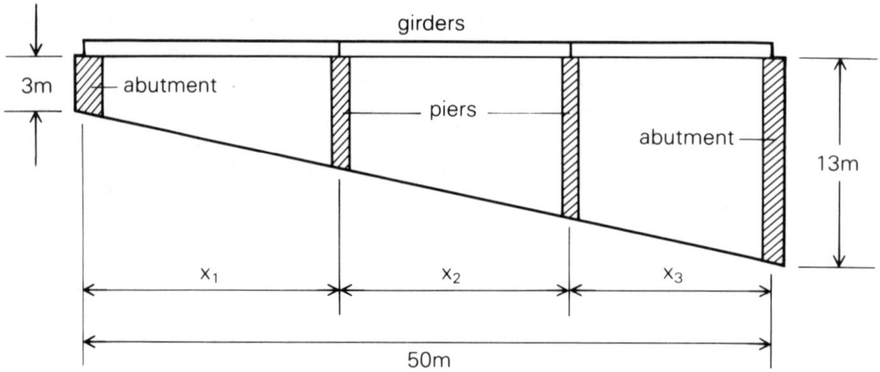

Figure 8.13 A three span bridge.

All miscellaneous costs such as the deck, railings, and footpaths can be ignored in the optimization study.

Compare the total cost of this design with the cost of the alternative using equal spans x_1, x_2, and x_3.

8.2 Find the minimum-cost dimensions for a cylindrical rainwater tank of volume 100 m^3. The material for the walls costs twice as much as the material for the ends. Assume that the ends must be cut out of square sheets of metal and the offcuts scrapped.

8.3 Portion of a trunk main under steady state conditions is shown in Figure 8.14. It is desired to find the minimum-cost diameters d_1, d_2, and d_3 such that the head loss from A to D does not exceed 30 m. The head loss in any section of pipe is given by the Darcy Weisbach formula:

$$H = \frac{f\,l}{d}\left(\frac{v^2}{2g}\right)$$

where H is the head loss (m), f is the friction factor, l is the length of pipe (m), d is the diameter of pipe (m), v is the velocity of flow in pipe (m/s), and g is the acceleration due to gravity (9.81 m^2/s).

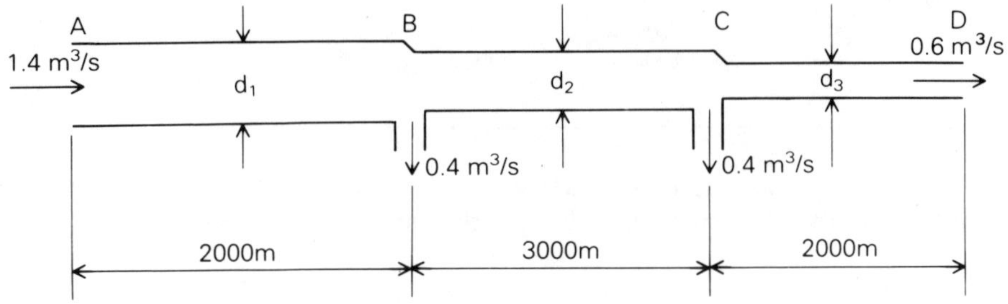

Figure 8.14 A trunk main.

In this case $f = 0.02$. Ignore minor head losses due to contractions, bends, etc. The cost of pipe (in \$/m length) is given by

$$c = 100\,d^{1.5}$$

Solve the problem using the Lagrange multiplier technique.

8.4 An optimal policy is required to maximize the net daily benefits due to the operation of a reservoir from which water may be extracted for both power generation and irrigation purposes.

Each 1000 m^3 of water released through the power station results in the generation of 150 kWh of electrical energy, with all of the water passing from the turbines into the river downstream of the dam. The net benefit to the community of power generation at this site is 10 c/kWh for the first 3000 kWh, 6 c/kWh for the next 3000 kWh and 4 c/kWh thereafter (up to a maximum of 15 000 kWh).

Each 1000 m^3 of water released for irrigation purposes produces a net benefit to the community of \$10. In addition, 30% of the water released for irrigation is returned to the river downstream of the dam.

The total volume of water available for power generation and irrigation purposes is 100 000 m^3/day. In order to maintain water quality standards in the river downstream of the dam, a minimum flow of 10 000 m^3/day must be maintained.

Formulate the problem of finding the daily volumes of water to be used for power generation and irrigation as an LP problem using the separable programming technique.

8.5 Rework Problem 7.7 assuming that the cost of waste removal is not constant at each site but is given in Table 8.5

Table 8.5 Costs of waste removal.

Capacity of plant (pollutant removal kg/day)	Cost ($\$\times 10^3$) of installing a plant at	
	Industry 1	Industry 2
0	0	0
500	100	120
1000	250	300
1400	—	500

Linear interpolation may be used for intermediate values of plant capacity. Why is it possible to use separable programming for this problem?

8.6 A trucking company wants to transport goods from city A to city B at minimum cost. The cost of travel on each link (in dollars) is shown in Figure 8.15. Find the minimum cost route from A to B. (Note that it may be necessary to introduce dummy states in order to apply dynamic programming to this problem.)

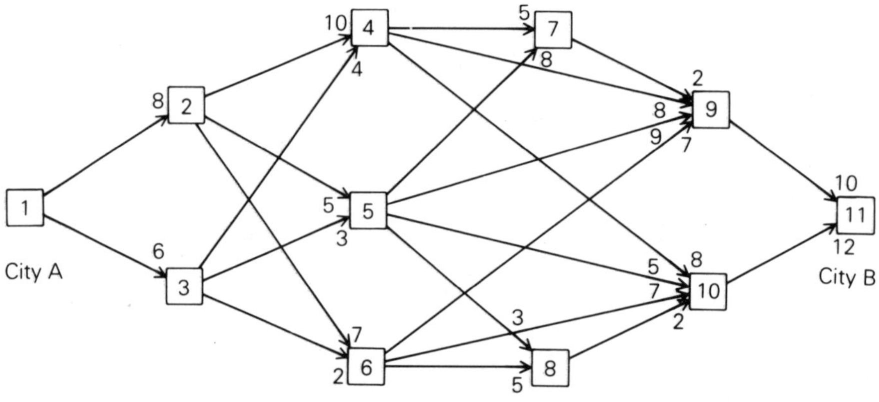

Figure 8.15 A highway network.

8.7 A quarry produces crushed rock for road-making. It can produce up to 5000 m^3 per month. The quarry's orders and cost of production per 1000 m^3 for the next four months are given in Table 8.6.

The quarry may store up to 4000 m^3 of crushed rock from one month to the next at a cost of \$50/1000 m^3.

Use dynamic programming to find the quarry's minimum cost production schedule over the four months assuming it starts and finishes with no crushed rock in storage.

Table 8.6 Orders and production costs.

Month	1	2	3	4
Orders (× 1000m^3)	2	4	4	3
Production cost for the first 1000m^3 ($)	500	550	600	550
Production cost for each additional 1000m^3 ($)	300	250	350	300

8.8 Water is pumped from a river into a reservoir and is used to supply water to a nearby city. The maximum volume of water which can be pumped into the reservoir in any month is 30 000 ML.

It is desired to determine the minimum-cost pumping schedule over the next four months so as to meet the known demands of the city. The demands and pumping costs are given in Table 8.7.

Table 8.7 Demands and pumping costs.

Month	Demand (Ml)	Cost ($M) of pumping a volume of: 0	10 000 ML	20 000 ML	30 000 ML
1	20 000	0.0	1.0	2.5	3.0
2	40 000	0.0	1.5	2.7	3.3
3	30 000	0.0	1.2	2.0	3.0
4	10 000	0.0	0.9	2.6	3.5

The reservoir contains a volume of 30 000 ML at the start of the first month and must contain at least 20 000 ML at the end of the fourth month. Evaporation and seepage losses are negligible.

Using dynamic programming and discrete volumetric units of 10 000 ML find the minimum-cost set of monthly volumes to be pumped to the reservoir over the period.

9 Engineering decision-making

Decision-making is central to engineering work. A simple introduction to mathematical decision theory is presented in this chapter.

Decision theory deals with simple, precisely defined (and therefore rather idealized) problems of choice. Real engineering decision situations are usually quite complex, and do not always lend themselves to the simplification needed to achieve one of the idealized situations for which a standard analysis is available. Nevertheless, the concepts of decision analysis provide a useful framework from which real problems may be approached. In particular, the identification of the alternative courses of action available, and then the systematic evaluation of possible outcomes and consequences, are very important first steps in dealing with any real decision problem.

9.1 Introduction

Decision-making is essentially a process whereby one course of action is chosen from a range of alternatives. Decision-making is an essential part of all engineering work, and indeed of professional work generally. It is also an inherent part of everyday life.

In a typical day we make innumerable personal decisions. In most cases the consequences are not of lasting significance, and so the decisions are of minor importance. Such routine decision-making is undertaken almost automatically, with little conscious thought or planning. For example, decisions are made in the morning when we decide to get up, choose which clothes to wear, decide what to eat for breakfast and how to travel to work.

Less frequently, we make personal decisions which are of great importance to us, perhaps because of their long-term effects on our life or because of the high cost involved. An important personal decision, such as to purchase a car or a house, or to change employment, is usually made with care and after much consideration. A commonsense procedure for such decision-making is:

- identify the alternative courses of action that are available
- determine the consequences of each alternative
- compare and evaluate the consequences on the basis of some suitable criterion such as cost or personal preference, and hence choose the course of action which leads to the most favourable consequences

Of course, important personal decisions are not always made in a completely commonsense, rational manner. For example, the decision to get married is a case where rational considerations may take a minor part in the decision process. Non-rational influences, such as current fashion, can also play an important role in much personal decision-making and indeed in 'professional' decision-making.

In Chapter 2 we have already seen how the identification of alternatives and then their evaluation are essential steps in the engineering methodology. That methodology is in fact a form of decision-making. Certainly decision-making is an essential part of the overall process of engineering planning and design. Decision-making is also an essential part of the more detailed stages of engineering work.

For example, the planning, design, and construction of a new airport for an urban area involves engineering decision-making at all stages. The initial choice of a site from various alternatives is of great importance to the inhabitants of the potential sites, as well as to local industries and to the many people involved in the operation of the airport. The social and political implications of decisions regarding airport siting will be familiar to most people from actual cases in cities as far apart as London, Tokyo, and Sydney. At the detailed level of airport design, numerous technical decisions also have to be made, for example in relation to the design and construction of pavements, buildings, access roads, and other facilities for aircraft, passengers, and support activities. Although the technical decisions at the detailed stages of design and construction do not normally lead to public interest or controversy, they can involve the design engineers in difficult technical decision-making.

Decision theory and decision analysis

Good decision-making is of crucial importance not only in engineering work but also in a wide spectrum of professional activities, including business administration and management. The processes of decision-making have consequently been studied intensively, especially in the last thirty years, in an attempt to develop reliable criteria and procedures for decision-making. A formal body of knowledge has been developed to deal with a range of decision situations. In the main, decision theory covers simple and rather idealized situations. Nevertheless, the methods of analysis have practical applications. Even when the theoretical methods of analysis are not directly applicable to the complex situations which occur in engineering practice, they can provide a useful insight into the appropriate decision-making procedures.

Decision models

In the study of decision-making processes it is useful to consider a variety of categories for which theoretical models have been developed.

ONE-OFF AND MANY-OFF DECISIONS

For purposes of analysis it is important to distinguish between one-off and many-off decisions. A **one-off** decision is made only once and involves a unique set of circumstances which will not be repeated. An example is the siting of a large bridge. A **many-off** decision is in some sense repeated a large number of times. For example, the design of a new fuel injection system for an automobile should lead to the production of many thousands of the item. In this sense, the design decision is many-off.

SINGLE-STAGE AND MULTI-STAGE DECISION PROCESSES

In a very simple decision situation it may only be necessary to make one isolated or **single-stage** decision, as for example the choice of a micro-computer for use in a small design office. In other circumstances, a sequence of inter-related decisions, i.e. a multi-stage decision process, is required. Choosing 'heads' or 'tails' prior to the toss of a coin is a single-stage decision; the choice of a sequence of moves in a game of chess is a multi-stage decision sequence. Choosing a detailed sequence of steps for the assembly of an electrical appliance in a factory assembly line is an example of a **multi-stage** decision process.

CERTAINTY, RISK, AND UNCERTAINTY OF OUTCOME

A further categorization involves the amount of information we have on the consequences of each of the possible courses of action. This can vary from full information concerning all possible actions and outcomes, through partial knowledge, to complete ignorance of the possible consequences of any course of action. The term 'certainty of outcome' is used to refer to the idealized situation where the decision-maker knows exactly what the outcomes of each possible course of action will be. 'Risk of outcome' describes the situation when the decision-maker knows what the possible outcomes are for each action, and also can estimate the relative likelihood of each outcome. 'Uncertainty of outcome' is used to describe the situation when the possible outcomes are known for each possible action, but nothing at all is known of the relative likelihood of occurrence of the outcomes.

DECISION-MAKING IN A COMPETITIVE SITUATION

A quite different situation arises if the outcome of any decision is affected by the actions of an intelligent adversary. For example, when a construction company tenders for a job in competition with other companies, there will be a range of prices it may choose for tendering, each price representing a possible course of action. However, the outcome of each course of action will be affected by the tender prices submitted by other tendering companies.

9.2 Single-stage decisions with risk of outcome

We commence our study of decision theory by considering a one-step decision situation with risk of outcome. It is assumed that a finite number of possible courses of action have been identified; they are denoted as a_1, a_2, . . ., a_i, . . ., a_m. For each course of action there is a set of possible outcomes. Which outcome will actually follow on from a course of action depends not on the decision-maker, but on chance. It is therefore convenient to imagine a number of possible states of nature, S_1, S_2, . . ., S_j, . . ., S_n, one of which will occur after the course of action has been decided on. The outcome of the decision is considered to depend on both the course of action chosen, a_i, and the state of nature which occurs, S_j, and is denoted as $O(a_i,S_j)$ or simply as O_{ij}. Each outcome is further assumed to have a value, or payoff in dollars. The payoff for O_{ij} is u_{ij}.

The case being considered is one with risk of outcome, which means that the decision-maker can estimate the relative probabilities of occurrence of the states of nature. These probabilities may be written as $p_1, p_2, . . ., p_j, . . . p_n$, where p_j is the probability $P(S_j)$.

As an example of a situation with risk of outcome, consider the design of a small, isolated farm building to resist high wind. The problem is to decide on an appropriate design wind velocity given that the maximum wind velocity in any time period is a highly variable quantity. If a very high value is chosen, the probability of failure of the building will be very low, but the cost will be high. On the other hand, choice of a low value for the design wind velocity leads to low cost but relatively high probability of failure. We might consider three possible actions as follows:

a_1: design the structure to resist a five-year wind W_5, i.e. a wind velocity which, on the average, is exceeded once in five years;
a_2: design the structure to resist a 25-year wind, W_{25};
a_3: design the structure to resist a 100-year wind W_{100}.

The estimated costs of the structure for the cases considered are \$10 000, \$14 000, and \$20 000 respectively; however, the cost of failure in each case would be \$40 000. The states of nature to be considered relate to the maximum wind velocity W_{max} which actually occurs during the design life of the structure. We consider the following possibilities:

S_1: the maximum wind velocity W_{max} does not exceed W_5;
S_2: W_{max} exceeds W_5 but does not reach W_{25};
S_3: W_{max} exceeds W_{25} but does not reach W_{100};
S_4: W_{max} exceeds W_{100}.

In this simple example, we ignore the possibility of more than one high wind occurring and causing several failures during the design life.

The relevant probabilities of the states of nature might be determined from meteorological data. The following values are used in this example: $P(S_1) = p_1 = 0.60; p_2 = 0.20; p_3 = 0.18; p_4 = 0.02$.

MATRIX REPRESENTATION

The numerical data for a decision with risk can be presented conveniently in a payoff matrix and a probability matrix. These are shown for the above numerical example in Table 9.1, and for the general case in Table 9.2. The payoff matrix tabulates the various u_{ij} for a vertical listing of the actions a_i and a horizontal listing of the states of nature, S_j.

Table 9.1 Decision with risk of outcome.

(a) Payoff matrix. Payoffs u_{ij} are in thousands of dollars; the negative sign indicates a cost.

Payoffs u_{ij}		States of nature S_1	S_2	S_3	S_4
courses	a_1	−10	−50	−50	−50
of	a_2	−14	−14	−54	−54
action	a_3	−20	−20	−20	−60

(b) Probability matrix

Probabilities p_{ij}		States of nature S_1	S_2	S_3	S_4	$\sum_j p_{ij}$
courses	a_1	0.60	0.20	0.18	0.02	1.00
of	a_2	0.60	0.20	0.18	0.02	1.00
action	a_3	0.60	0.20	0.18	0.02	1.00

In the introductory numerical example the probability of occurrence of any state of nature is unaffected by the specific action chosen, and it is appropriate to use a single subscript for the probability, i.e. p_j for the probability of occurrence of S_j. Nevertheless, it is convenient to use double subscripts for the probability and write p_{ij} to represent the probability of the jth outcome following on from action a_i. The double subscript does no harm in the present example, and it becomes necessary in other more complex situations. The probability matrix is therefore shown in Table 9.2(b) with double subscripts. It is to be noted that the probabilities for any particular action sum to unity (i.e. the row sums are unity):

$$\sum_{j=1}^{n} p_{ij} = 1.0 \tag{9.1}$$

EXPECTED VALUE DECISION CRITERION

In order to make a rational decision in the present situation, some appropriate criterion needs to be identified. The **expected value criterion**, which is frequently used in cases involving risk of outcome, states that the best course of action is the one which has the highest expected value (or least expected

Table 9.2 Payoff matrix and probability matrix.

(a) Payoff matrix.

Payoffs u_{ij}		States of nature S_1		S_j		S_n	$EV(a_i)$
	a_1	u_{11}	...	u_{1j}	...	u_{1n}	
	.	.		.		.	
courses	.	.		.		.	
	.	.		.		.	
of	a_i	u_{i1}	...	u_{ij}	...	u_{in}	$\sum_j u_{ij} p_{ij}$
	.	.		.		.	
action	.	.		.		.	
	.	.		.		.	
	a_m	u_{m1}	...	u_{mj}	...	u_{mn}	

(b) Probability matrix

Probabilities p_{ij}		States of nature S_1		S_j		S_n	$\sum_j p_{ij}$
	a_1	p_{11}	...	p_{1j}	...	p_{1n}	1.00
	.	.		.		.	
courses	.	.		.		.	
	.	.		.		.	
of	a_i	p_{i1}	...	p_{ij}	...	p_{in}	1.00
	.	.		.		.	
action	.	.		.		.	
	.	.		.		.	
	a_m	p_{m1}	...	p_{mj}	...	p_{mn}	1.00

cost) associated with it. For any course of action a_i, the expected value is defined as

$$EV(a_i) = \sum_{j=1}^{n} u_{ij} p_{ij} \tag{9.2}$$

For the numerical example above, the expected values of the three courses of action are:

$$EV(a_1) = 0.6(-10) + 0.2(-50) + 0.18(-50) + 0.02(-50) = -26.0$$

$$EV(a_2) = 0.6(-14) + 0.2(-14) + 0.18(-54) + 0.02(-54) = -22.0$$

$$EV(a_3) = 0.6(-20) + 0.2(-20) + 0.18(-20) + 0.02(-60) = -20.8$$

which suggests that action a_3 is the best.

Under which circumstances is it appropriate to use the expected value criterion? If the decision is many-off, then expected value represents the long-term return and the criterion is highly appropriate. Examples of many-off decisions would be a mass purchase of microcomputers, the choice of precast cladding units for a tall building, or periodic replacement of electronic components with a known probability of failure.

If a large number of one-off decisions of a similar type are being made, then it is appropriate to use the expected value criterion for each one-off decision because the results will 'average out' and the end result will be the same as for a many-off decision. This situation often applies to decisions made in a large organization, such as a government department. Consider, for example, the decision by a government department whether or not to take out insurance on its many vehicles. On average, the insurance premium will exceed the cost of damage each year due to administration costs and profits taken by the insurance company. The department would normally be better off not to insure, but to bear the risk itself.

For the expected value criterion to be appropriate, the cost or benefit of the decision needs to be relatively small in relation to the overall operations of the decision-making body. When the possible losses and gains from one decision are of the same order of magnitude as the operating budget of the decision-maker, then one or two chance outcomes could result in financial difficulty or even ruin. In such circumstances, few people would be willing to use the expected value criterion. Before looking at a more general criterion, we shall first consider an alternative method for representing the simple decision situation with risk of outcome.

DECISION TREE REPRESENTATION

Provided the number of courses of action, m, and the number of possible states of nature or outcomes, n, are not large, a useful visual representation of a decision situation is achieved by means of a decision tree. A decision tree for the numerical problem in Table 9.1 is shown in Figure 9.1. The decision node, or point at which the decision-maker makes a decision, is represented by the square box. The possible courses of action are represented by action lines, in this case three, emanating from the decision node. Each action line terminates in a round node, which might be thought of as representing a chance event, or a decision by Nature. The alternative outcomes for each action are represented by the lines emanating from the round nodes. Values of probability are noted on each outcome branch. The two boxes at the extremity of each decision path (a_i,S_j) contain the values for the payoff u_{ij}, and the expected value $p_{ij}u_{ij}$. As indicated by Equation (9.2), the expected value of a particular course of action is obtained by summing the expected values at the end of the associated branches.

The decision tree is especially useful in simple situations. It can be used to advantage when the outcomes depend on the course of action chosen or when the concept of state of nature is inappropriate.

In Figure 9.2 a decision tree is used to represent a problem in the field of

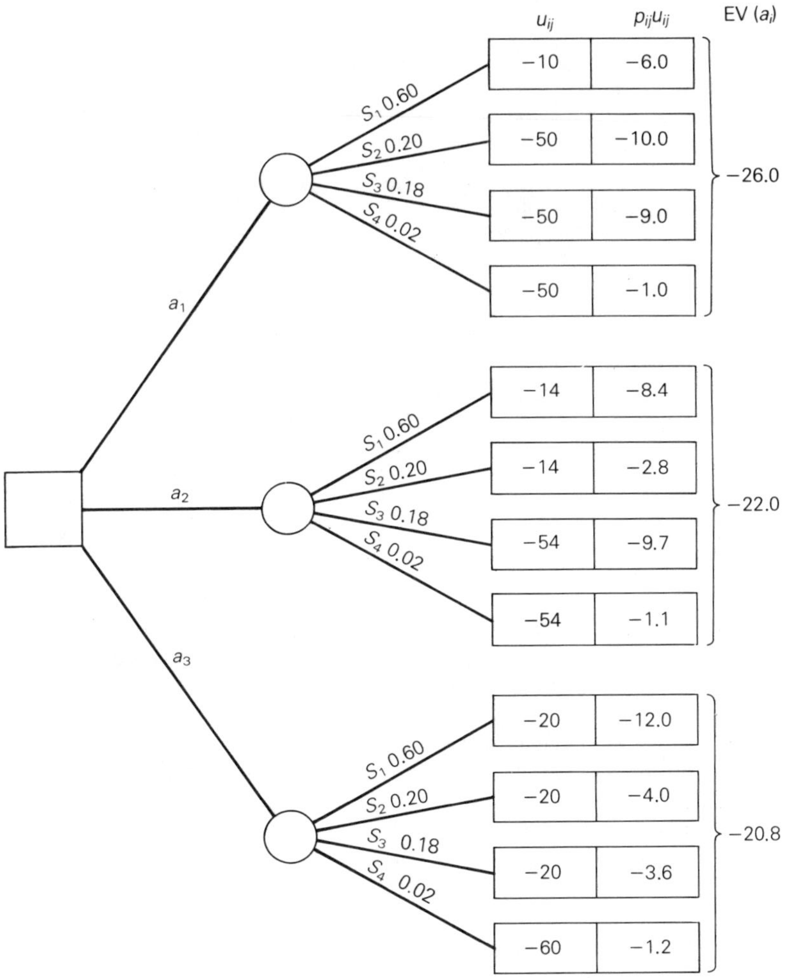

Figure 9.1 Decision tree.

structural engineering. The problem is to decide whether a small bridge, which is showing symptoms of deterioration and distress, is safe or not. Three courses of action are contemplated:

- a_1: condemn the bridge and replace it (in the belief that it is in danger of collapse)
- a_2: repair the bridge (in the belief that it is defective, but repairable at reasonable cost)
- a_3: do nothing (in the belief that the symptoms are not indicative of any serious defect)

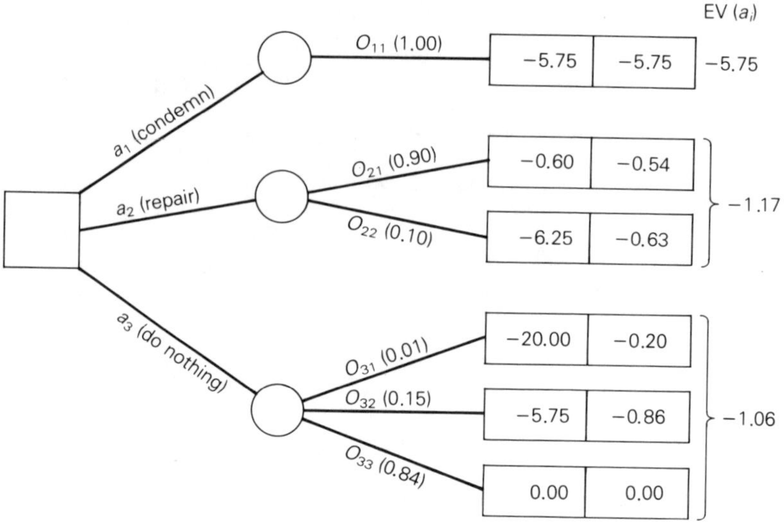

Figure 9.2 Decision tree for bridge repair/replacement.

In this problem it is convenient to deal directly with outcomes, rather than with 'states of nature'. We consider the following possible outcomes and associated costs (in millions of dollars):

O_{11}:	the bridge is successfully replaced;	$u_{11} = -\$5.75$
O_{21}:	the repairs are satisfactory;	$u_{21} = -\$0.60$
O_{22}:	the repairs are unsatisfactory;	$u_{22} = -\$6.25$
O_{31}:	the bridge collapses;	$u_{31} = -\$20.0$
O_{32}:	the bridge becomes unserviceable;	$u_{32} = -\$5.75$
O_{33}:	the bridge remains satisfactory;	$u_{33} = -\$0.0$

In the payoffs, the negative sign indicates a cost, i.e. a loss and not a gain.

In Figure 9.2 the expected value of any action a_i is obtained by adding together the expected values of all branches associated with that action line. In this example, a_3 has the lowest expected cost, even though one of the possible outcomes, O_{31}, is collapse of the bridge. If collapse involves risk of injury or loss of life, then clearly an equivalent monetary cost of life and limb, calculated in some suitable manner, has to be included in the estimate of u_{31}. The concept of utility, which will be introduced shortly, provides a useful basis for handling outcomes with non-monetary implications. The matter of non-monetary outcomes is also discussed in Chapter 6.

It will be appreciated that the 'best' course of action depends on the probabilities p_{ij}. These values are estimated on the basis of the available information, which is likely to be inadequate and subjective in nature. In such circumstances it may be useful to recalculate the decision analysis several times with varying probability values in order to test the sensitivity of the result to the input data.

EQUIVALENCE OF MATRIX AND DECISION TREE REPRESENTATIONS

In the example in Figure 9.2, the number of outcomes varies from action to action and the decision tree provides a convenient method for analysing the problem. Nevertheless, we could have used the matrix method. Indeed it is always possible to develop an equivalent matrix formulation by introducing an appropriate number of 'dummy' states of nature. If necessary, the number of states, n, can be equal to the total number of distinct outcomes, or branches, in the tree. In Table 9.3 a matrix treatment of the problem of Figure 9.2 is given. Six states of nature are used, one for each outcome. It will be noted that this is an example where the double subscript for probabilities is required.

Table 9.3 Equivalent matrix representation.

(a) Payoff matrix u_{ij}

u_{ij}		States of nature					
		S_1	S_2	S_3	S_4	S_5	S_6
courses	a_1	$u(0_{11})$	0.0	0.0	0.0	0.0	0.0
of	a_2	0.0	$u(0_{21})$	$u(0_{22})$	0.0	0.0	0.0
action	a_3	0.0	0.0	0.0	$u(0_{31})$	$u(0_{31})$	$u(0_{33})$

(b) Probabilities

p_{ij}		States of nature					
		S_1	S_2	S_3	S_4	S_5	S_6
courses	a_1	1.00	0.0	0.0	0.0	0.0	0.0
of	a_2	0.0	0.90	0.10	0.0	0.0	0.0
action	a_3	0.0	0.0	0.0	0.01	0.15	0.84

(c) Expected values

$$\mathrm{EV}(a_1) = 1.00\,(-5.75) = -5.75$$
$$\mathrm{EV}(a_2) = 0.90\,(-0.60) + 0.10\,(-6.25) = -1.17$$
$$\mathrm{EV}(a_3) = 0.01\,(-20.0) + 0.15\,(-5.75) + 0.84\,(0) = -1.06$$

RISK AVERSION

The expected value criterion is appropriate for use when many-off decisions are being made, or when many one-off decisions of a similar type are undertaken. The implication is that monetary profits and costs are a suitable measure of value. However, expected value is not a suitable criterion in situations where the costs of the outcomes are large in relation to the financial backing of the decision-maker. This is because individuals and organizations often display in decision-making a phenomenon known as **risk aversion**.

To illustrate risk aversion, we consider a simple game which involves the spin of a 'chocolate wheel' (Figure 9.3). The rules of the game are as follows: If the pointer indicates black when the wheel stops turning, the player wins. In this case the win is not a box of chocolates, but \$1000. On the other hand, if white is indicated, nothing is won. Furthermore, it costs \$$X$ to purchase a ticket for each play of the game. The probability of black being the outcome of the game is 0.5. The question of interest is for what values of X will a decision-maker choose to play this game. We analyse the decision situation in Figure 9.3(b), where there are two possible courses of action: to play, a_1, or not to play, a_2.

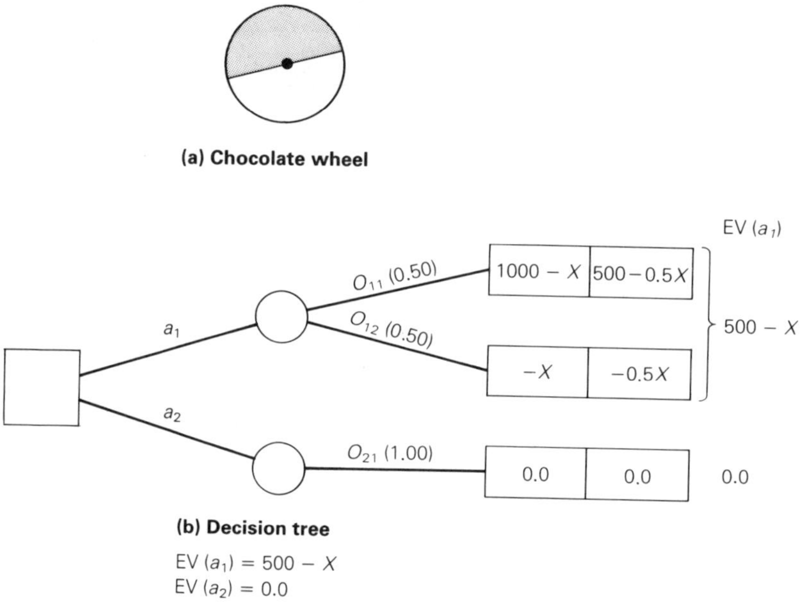

Figure 9.3 Risk aversion.

According to the expected value criterion, the correct decision is to play, provided that X does not exceed \$500. Nevertheless, few people would be willing to pay this much for a 50 per cent chance of winning \$1000. They would consider that their financial resources are inadequate to cover a one-off loss of \$500. Even with substantial financial resources, many people would only be willing to play this game repeatedly if X were much less than \$500.

The term 'risk aversion' is used to describe the situation where decision-makers are unwilling to accept the expected value criterion for decision-making because of their aversion to risking the loss of considerable sums of money. An example of risk aversion in everyday life occurs whenever individuals and organizations buy insurance. The expectation is that the insurance company will profit on the agreement; nevertheless, individuals happily insure houses and automobiles at relatively low cost, in order to cover the unlikely event of large-scale property loss. A simple, idealized decision situation comparable with the purchase of insurance is shown in Figure 9.4. In this case the decision-maker has two choices, either to pay \$150 to insure an investment worth \$100 000 for a year, or to accept a very low risk of losing that investment during the year. In such situations, most people choose action a_1 even though the expected return is less than for action a_2.

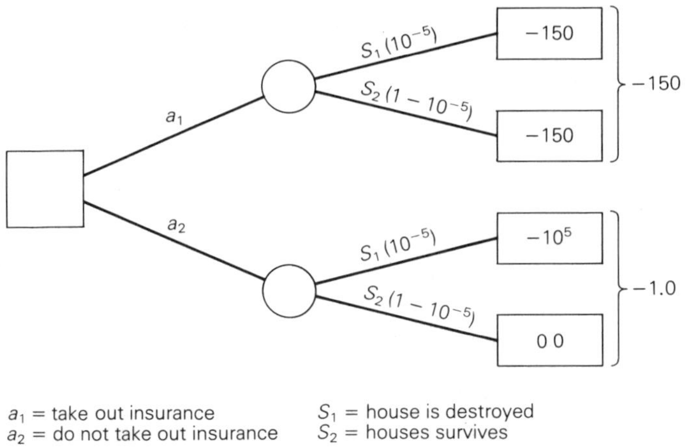

Figure 9.4 Insurance and risk aversion.

RISK ACCEPTANCE

It is interesting that the same individuals who display risk aversion in some circumstances, may in other circumstances display risk acceptance. The term **risk acceptance** refers to the willingness of a decision-maker to undertake a course of action when the expected value criterion suggests that the action should not be taken.

A common example of risk acceptance is the willingness of many individuals to gamble. A simple gambling situation may be conceived in terms of the chocolate wheel previously referred to (see Figure 9.5). This time, the probability of black coming up is very small, say 10^{-6}. However, the cost of purchasing a ticket is now only \$5 and the win is \$10 000. Many people are happy to gamble \$5 on lotteries of this kind. The expected value of this action is very close to −\$5, and the expected value criterion suggests that the correct action is not to play.

UTILITY

It is possible to take proper account of both aversion to and acceptance of risk by introducing the concept of **utility**. By using a utility measure of the value of each outcome in place of dollar values it is possible to apply the

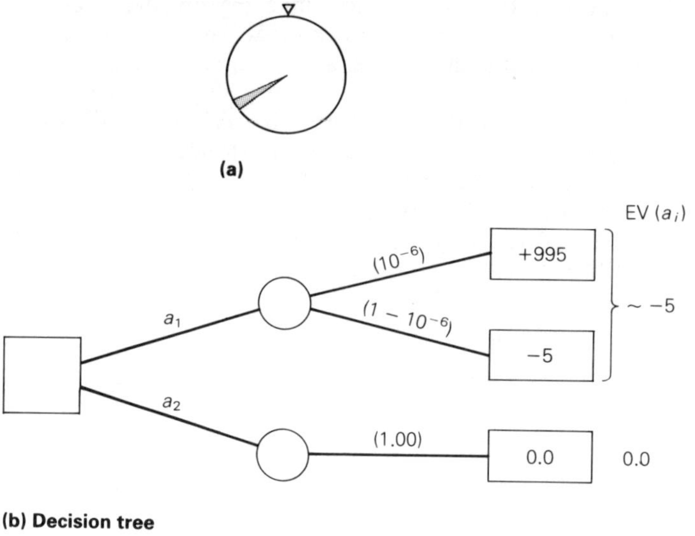

Figure 9.5 Lottery: risk acceptance.

expected value criterion to obtain a valid decision criterion. It is very important to recognize that decision-making is a subjective activity, so that the correct decision for one individual may not apply to someone else with a different attitude to risk taking. Nevertheless, a consistent and rational approach can be developed by means of the concept of utility, as introduced by von Neumann and Morgenstern (1947).

The essential step is the construction of a **utility function** which transforms the monetary value of the outcomes of a set of possible actions into non-dimensional units of utility. This function is subjective, in that it applies only to the decision-maker for whom it is set up. It may also apply only to the particular decision situation currently being considered. In addition, the attitude to risk of a decision-maker may change with time, as a result of changes in financial and other circumstances.

To construct the utility function for a specific situation, it is necessary to inspect all the possible outcomes and identify the most favourable and the least favourable. These are denoted as O^* and O_*, respectively. An appropriate range of values for the units of utility is then chosen. This may for example be from 0 to 100 units or 0 to 1 unit, from −100 units to 0, or perhaps from −50 to +50 units, depending on the physical circumstances. The lowest value is allocated to outcome O_* and the highest to O^*. Thus, using a range from 0 to 100, we express the extreme utilities as $U(O_*) = 0$ and $U(O^*) = 100$.

A game, or lottery, is now considered by the decision-maker (Figure 9.6) in which two courses of action are possible, a_1 and a_2. The possible outcomes of a_1 are O^* and O_* with probabilities of p and $(1 - p)$, respectively. The value of p will be considered shortly. Action a_2 has a single outcome O_i, which is intermediate between O^* and O_* and by definition must have a utility value between $U(O^*)$ and $U(O_*)$.

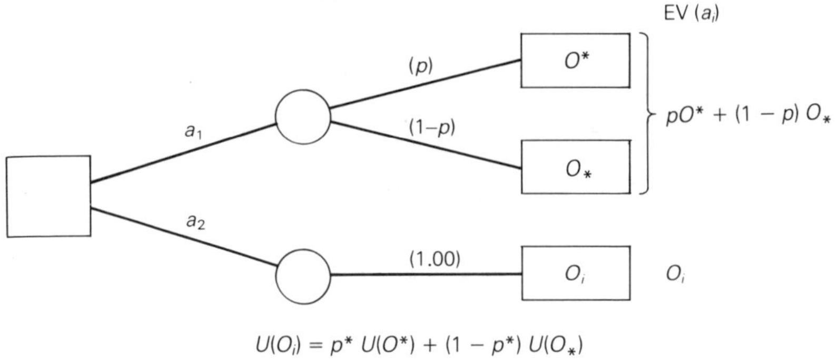

Figure 9.6 Lottery for the evaluation of utility.

For this situation, shown in Figure 9.6, the preference for a_1 or a_2 depends on the probability p. Clearly if $p = 1.0$ then a_1 is preferable and if $p = 0.0$ then a_2 is preferred. It is necessary for the decision-maker to find the probability value p^* for which the two actions are equally acceptable. This can be done by beginning with $p = 1.0$ and progressively decreasing the value and interrogating the decision-maker as to the acceptability of a_1. In the case of an organizational decision, an individual representative or a group of individuals will obviously be consulted in this way. When two actions are equally acceptable we define their expected utilities to be equal. Thus, when the probability p^* has been determined the utility of O_i is calculated as

$$U(O_i) = p^*U(O^*) + (1 - p^*)U(O_*) \tag{9.3}$$

A sequence of values for O_i are considered in turn and their utilities evaluated so that a graph can be constructed for transforming any outcome into an appropriate utility.

Example 9.1 Consider a small engineering software company with little financial backup, and facing a business situation which involves possible losses as large as \$15 000 and possible profits as high as \$30 000. The general decision situation is shown as a decision tree in Figure 9.7(a) which suggests that action a_1 is most appropriate if the analysis is carried out purely in monetary terms. With $O_* = -\$15\,000$ and $O^* = \$30\,000$, the utility lottery appears as in Figure 9.7(b) for the intermediate case $O_i = 0$. We see that a high value of p^* would be needed to tempt the firm to take action A_1. A value of 0.99 would almost certainly be acceptable. A lower value of 0.9 might well be acceptable. The cut-off is here assumed to be 0.8. With the utilities of O_* and O^* defined to be 0 and 100, respectively, the resulting utility value of \$0 is found from Equation (9.3) to be $U(0) = 80$. Following a similar procedure for a loss of \$10 000 and a gain of \$15 000, the function shown in Figure 9.7(c) is obtained.

Using utilities in place of dollar outcomes, the expected value calculations are as shown on the decision tree in Figure 9.7(d) and action a_2 now becomes the best one.

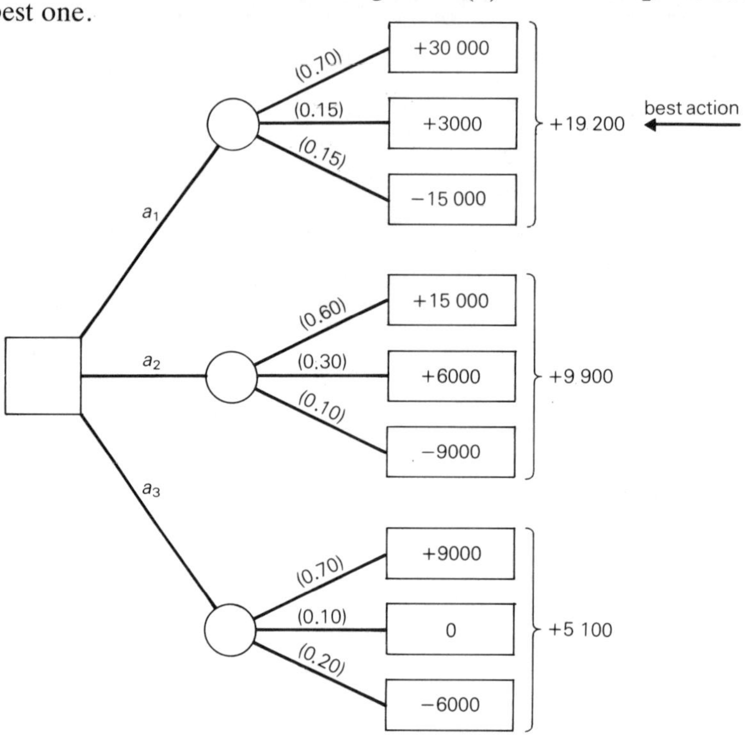

(a) Expected value decision in terms of dollars

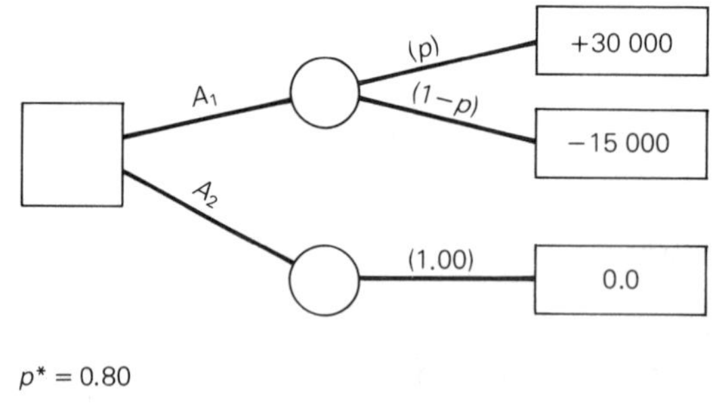

(b) Lottery to determine utility of $0.0

Figure 9.7 Decision analysis using utilities.

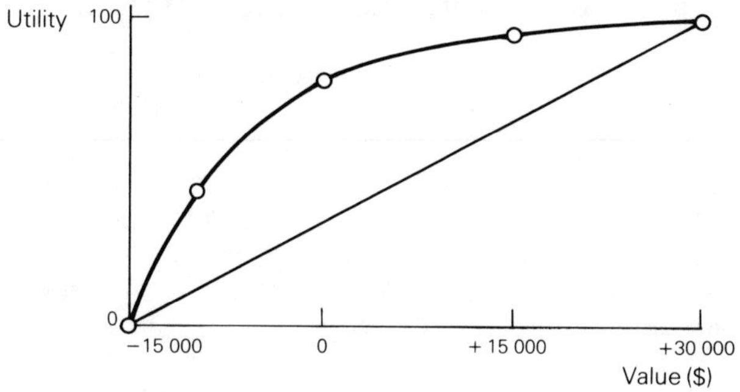

(c) Utility function

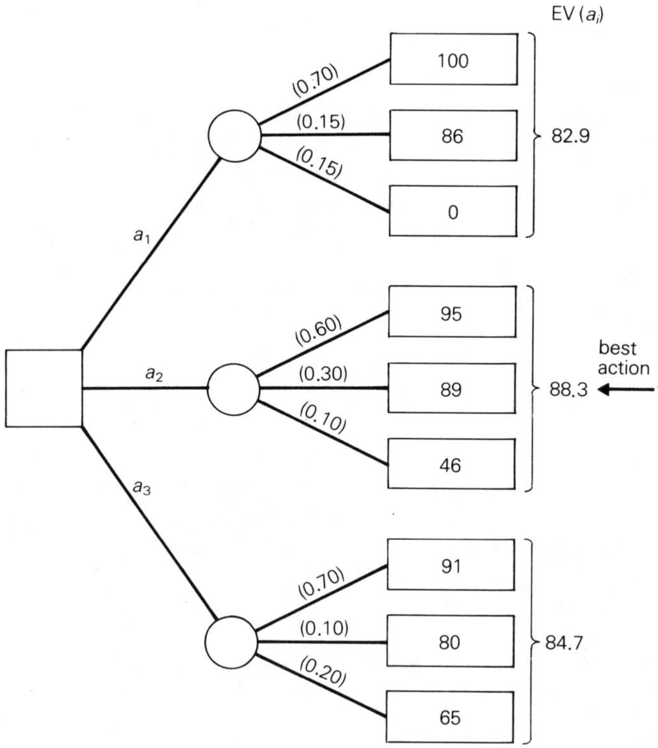

(d) Decision tree with utilities

Figure 9.7 ***contd.***

9.3 Decisions with certainty and uncertainty of outcome

The normal decision situation involving risk of outcome is one in which the decision-maker can estimate the probabilities of the possible outcomes of any particular course of action. The decision-maker is in a situation of having incomplete information. Two limiting cases are also of interest. The first arises when full information is available, so that the decision-maker is certain of the outcome of each course of action; the second applies when there is insufficient information to make any estimate of the relative probabilities of possible outcomes and is referred to as decision with uncertainty of outcome. We consider these cases in turn.

DECISION WITH CERTAINTY OF OUTCOME

In some situations the decision-maker may have a very good idea of the outcome which will follow on from each of the possible courses of action. If the outcomes can be evaluated in financial terms, then the decision process is simple and straightforward.

In engineering practice, it is not usually the case that outcomes can be predicted with near certainty. Even then, conditions are often too complex to allow decisions to be taken purely on the basis of monetary value. For example, we have already seen how social benefits and disbenefits need to be considered when we choose from alternative plans for an engineering project. In engineering decision-making, non-monetary aspects of evaluation are usually important. The situation is then comparable with that of evaluating a multi-objective project, as already discussed in Chapter 6.

An example of a situation which approaches the certainty-of-outcome model might be the choice of a computer facility for an engineering design office. As the alternative 'candidate' systems are offered by the various suppliers with full specifications, guarantees of performance, and maintenance agreements, it might be reasonable to assume the situation is one of certainty of outcome. Nevertheless, the choice may not be easy because the alternative systems will almost certainly reach varying levels of performance. Some of the important factors to be considered might include:

- initial cost
- recurrent costs (maintenance and software agreements)
- reliability, and time lost due to malfunction
- percentage of jobs too large to be handled efficiently
- quality and speed of support service
- range of software available

Clearly these factors are interdependent. Good performance in one area might mean reduced performance in another. Thus, low initial cost may well mean poor reliability and lack of software.

A commonsense approach to decision-making in such situations is to choose the most important or prime factor as the basis for the comparison, and then try to trade-off relative levels of performance in the other secondary factors by adjustments to the level of this prime factor. In the present example, initial cost might be chosen for the comparison. Recurrent costs can be evaluated in terms of present value, using the methods explained in Chapter 5, and equivalent financial values can then be placed on items such as reliability, downtime, and software availability.

An alternative approach, sometimes referred to as **satisficing**, is to choose one main parameter as the basis for the comparison, and then to set minimum acceptable standards for all other factors of importance. Any course of action which does not achieve minimum standards for all secondary factors is eliminated. The final decision is then based solely on the main factor. In the choice of the computer system, taking initial cost as the prime factor, we might use minimum standards such as the following:

- **service time:** within three hours of call during office hours
- **downtime:** not more than five per cent
- **recurrent costs:** not in excess of 10 per cent of the initial cost for the first five years of operation
- **software:** FORTRAN, BASIC, specific engineering analysis packages, word processing, statistics package, spreadsheet, business package

Before we leave the present topic it should be pointed out that the assumption of certainty of outcome is often used but rarely appropriate in engineering planning, design, and decision-making. Although unpalatable initially, both to the engineer and to the community, the concept of a finite but small risk of failure is gradually becoming accepted as the appropriate basis for engineering design work. The alternative of absolute safety is likely to be either physically unattainable or exorbitant in cost. Design procedures and codes of practice are increasingly being developed with the objective of achieving an acceptably low probability of failure. Replacement of absolute safety by a low but acceptable risk of failure as the goal of engineering planning and design leads to significant savings, a more effective use of national resources, and hence to significant social benefits (Asplund 1968).

DECISION WITH UNCERTAINTY OF OUTCOME

The term 'structured uncertainty of outcome' is used to describe situations in which the decision-maker can identify the possible outcomes of each course of action, but has insufficient information to assign probabilties to those alternatives. Although a precise model exists for this situation, it tends to be a rather artificial one, at least in engineering applications. Nevertheless, the model has received considerable attention and various criteria have been proposed for making decisions in the case of structured uncertainty (Taha 1982).

According to Laplace's Principle of Insufficient Reason, all outcomes may be assumed to be equally likely if no other information is available. The decision can then be treated as one with risk of outcome. A weakness in this approach is that spurious results will be obtained if additional outcomes are taken into account, no matter how unlikely they might be in reality. The approach leads to poor decision-making, and it is probably better to base decisions on intuition and past experience rather than on this assumption.

A **pessimistic** criterion assumes that the decision-maker should expect the worst and plan accordingly. This implies that decisions should be made in such a way that either the maximum possible loss is minimized or that the minimum possible gain is assured. Such an approach is indeed appropriate when the decision situation involves intelligent adversaries who can influence the choice of outcomes. This might for example be the case when several parties are competing for an engineering project. A detailed discussion of decision-making procedures in competitive situations is deferred to Section 9.6.

An **optimistic** criterion can also be postulated, which leads to decisions aimed at maximizing the maximum gain. Such an approach implies that Nature is basically benevolent. Although there are interesting situations in engineering analysis where it might be argued that Nature is indeed benevolent, this is clearly not an acceptable general basis for engineering decision-making.

A further criterion takes account of the attitude of the decision-maker after the outcome of the decision becomes known, and is based on the idea of regret. If the best possible action has been taken for the actual state of nature, then the decision-maker will have no regrets. For any state of nature a measure of regret for any action is obtained by subtracting the utility of that action from the utility of the best possible action for the state of nature being considered. The proposed criterion is then to minimize the maximum regret. While this approach may in some cases be useful in personal decision-making, it does not appear to be appropriate for engineering decision-making.

The present discussion has centred on a state of structured uncertainty. In practice, the decision-maker may well be in an even less advantageous position whereby not all of the possible outcomes will be known for any contemplated course of action. The term unstructured uncertainty describes the extreme situation when nothing is known about the possible outcomes of a course of action. No rational basis exists for decision-making at this stage, apart from expending time, effort, and money to acquire additional information.

Example 9.2 As an example of decision-making with uncertainty, we consider the payoff matrix in Table 9.1(a), ignoring the probability matrix in Table 9.1(b). It will be noted that the negative signs in this matrix indicate that we are dealing with costs and not gains.

Laplace criterion

If Laplace's criterion is used, a probability value $p_{ij} = 0.25$ is used throughout and we obtain:

$$\begin{aligned} \mathrm{EV}(a_1) &= 0.25\,(-10 - 50 - 50 - 50) = -40 \\ \mathrm{EV}(a_2) &= 0.25\,(-14 - 14 - 54 - 54) = -34 \\ \mathrm{EV}(a_3) &= 0.25\,(-20 - 20 - 20 - 60) = -30 \end{aligned}$$

This suggests that action a_3 should be chosen.

Pessimistic criterion
To apply the pessimistic criterion, we determine the minimum u_{ij} for each row in turn:

$$\begin{aligned} \text{for } i = 1, \min_j(u_{1j}) &= -50 \quad (j = 2,3,4) \\ i = 2, \min_j(u_{2j}) &= -54 \quad (j = 3,4) \\ i = 3, \min_j(u_{3j}) &= -60 \quad (j = 4) \end{aligned}$$

To maximize the minimum payoff we choose action a_1. In this example, this is equivalent to minimizing the maximum cost.

Regret criterion
To construct the regret matrix r_{ij} we identify the maximum payoff (or minimum cost) in each column in turn:

$$\text{for } j = 1, \max_i(u_{i1}) = -10 \quad (i = 1)$$

$$j = 2, \max_i(u_{i2}) = -14 \quad (i = 2)$$

$$j = 3, \max_i(u_{i3}) = -20 \quad (i = 3)$$

$$j = 4, \max_i(u_{i4}) = -50 \quad (i = 1)$$

The $(i-j)$th entry in the regret maxtrix is

$$r_{ij} = \max_i(u_{ij}) - u_{ij}; \qquad j = 1, \ldots, m$$

This gives

$$\left[r_{ij} \right] = \begin{bmatrix} 0 & 36 & 30 & 0 \\ 4 & 0 & 34 & 4 \\ 10 & 6 & 0 & 10 \end{bmatrix} \quad \begin{matrix} \text{max} \\ 36 \\ 34 \\ 10 \end{matrix}$$

To minimize the maximum regret, we choose action a_3.

BUYING INFORMATION
The decision situations we have to date considered, listed in decreasing order of information available to the decision maker, are as follows:

- certainty of outcome (full information on outcomes)
- risk of outcome
- structured uncertainty
- unstructured uncertainty (zero information on outcomes)

The acquisition of further information is always an important option to consider. The additional information may make it possible to move up one or two steps on the above scale. Even when the decision situation remains risky, additional information can improve decision-making by improving the probability estimates or even eliminating some of the less likely outcomes from consideration.

An example of the type of improvement achieved from additional information is shown in Table 9.4. Initially, the decision situation involves three possible states of nature with probabilities as in Table 9.4(a). The preferred action is a_3. With additional information, the unlikely state S_3 is eliminated altogether, and this in turn leads (Table 9.4(b)) to a change in expected costs and the choice of action a_1 rather than a_3.

Table 9.4 Effect of additional information.

(a) Initial decision situation

Payoff matrix		States of nature S_1	S_2	S_3	$EV(a_i)$	
Probabilities		0.30	0.60	0.10		
courses	a_1	5	7	−14	4.3	
of	a_2	6	5	−8	4.0	best
action	a_3	5	5	−1	4.4 ←	action

(b) With additional information

Payoff matrix		States of nature S_1	S_2	$EV(a_i)$	
Probabilities		0.20	0.80		
					best
courses	a_1	5	7	6.6 ←	action
of	a_2	6	5	5.2	
action	a_3	5	5	5.0	

In engineering decision-making, additional information may be acquired in various ways, for example through laboratory testing, pilot studies, theoretical analysis, and field trials. Although information gathering is inevitably costly, the value of improved decision-making may well more than compensate.

On the other hand, there are usually serious disadvantages in delaying engineering work too long and a balance often has to be struck between immediate action and excessive delay. This is illustrated in Figure 9.8 where C_1 represents the cost of postponing a decision and purchasing information, while C_2 is the expected added cost of a poor decision. As the postponement time increases, C_1 increases and C_2 decreases. The optimal postponement time, and the optimal amount of information to obtain, should be such that the sum of C_1 and C_2 is minimized (Figure 9.8(c)). In Figure 9.8 an implied assumption is that quantity of information is related to time taken to acquire the information. In some circumstances, it might be appropriate to consider directly the information available, i.e. replace 'time' in the horizontal scale by 'information'.

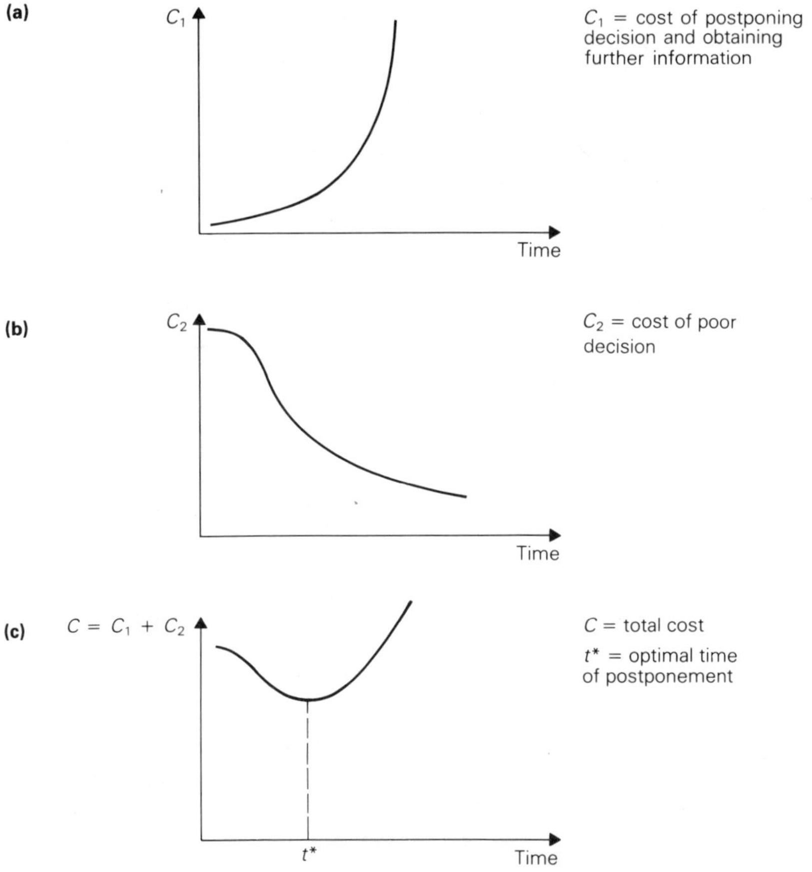

Figure 9.8 Obtaining additional information.

An interesting and important question is always whether it will be worth while to gather extra information or to make the decision immediately. The situation illustrated in Table 9.4 is shown as a decision tree in Figure 9.9. At an initial decision node, the choice between immediate decision-making (A_1) or gathering information (A_2) is made. A second stage of decision-making follows. A proper decision at the first node can only be made if the decision-maker already has a sufficiently good idea of the possible outcomes of the information-gathering exercise to estimate the expected cost of A_2. It should be noted that the cost of gathering information also increases the expected cost of A_2. In relation to Figure 9.8(c), action A_2 will not be chosen if the sum $C_1 + C_2$ continually increases with time and shows a minimum value at time zero.

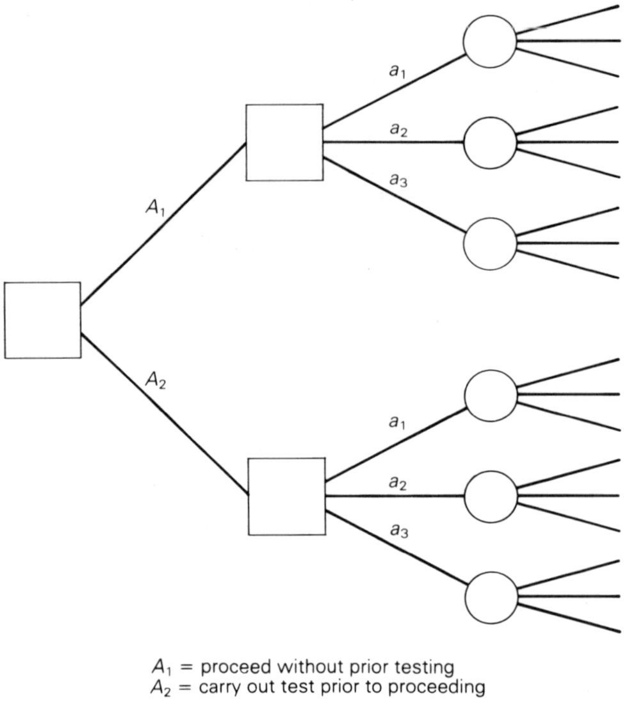

Figure 9.9 Preliminary decision whether or not to gather additional information.

The decision situation in Figure 9.9 is an example of a two-stage decision process. We go on in the next section to consider multi-stage processes. The question of how to use new information to obtain an improved estimate of probabilities of outcomes is discussed in Section 9.5.

9.4 Multi-stage decision processes

The methods of analysis previously considered have applied to a single-stage decision. Frequently, a series of sequential, inter-related decisions needs to be made. A simple two-stage decision situation was discussed at the end of Section 9.3. If in a multi-stage process the outcome from the nth decision becomes known before a decision has to be made at stage $(n+1)$ then it is possible to adjust the choice of actions to allow for the previous chance outcomes or states of nature. Nevertheless, an initial analysis of the entire process is needed in order to determine the best course of action to choose at the first step.

In the case where each step is undertaken with risk of outcome, the process can in principle be represented by a decision tree. In Figure 9.10 Stage 1 of the process is represented by a decision node with possible actions a_{1i}, each with subsequent outcomes O_{1ij}. The outcome of Stage 1 is some sequence (a_{1i}, O_{1ij}). Stage 2 follows on with a decision node and possible actions a_{2i}. At the kth stage, the available actions are a_{ki}, with possible outcomes O_{kij}. A particular decision sequence is defined by the full listing of index values of the sequence, $(a_{1i}, O_{1ij}; a_{2i}, O_{2ij}; \ldots a_{ki}, O_{kij})$. It may of course be more convenient to work with states of nature, S_{kj}, in lieu of outcomes O_{kij}.

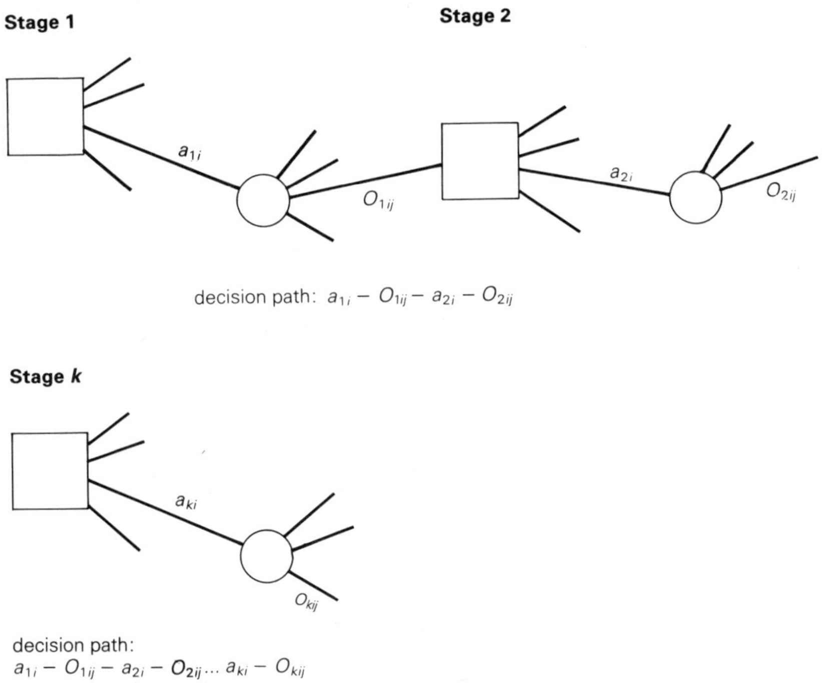

Figure 9.10 Multi-stage decision process.

The procedure for analysing a multi-stage decision process follows from the simple observation that, irrespective of the prior decisions and outcomes, the best course of action to choose at any stage of the process is the one which maximizes the total payoff for all future steps. To begin the analysis it is therefore appropriate to go to the last stage of the process and consider in turn each of the last-stage decision nodes in the tree. At each node in turn, the expected value of each course of action is calculated from the values entered at the tips of the associated state-of-nature branches and the probabilities of these branches. The course of action with the best expected value is thus identified, and this expected value is noted on the decision node.

When all last-stage decision nodes have been analysed, it is possible to move back to the second-last decision nodes and analyse these as one-step decisions. The expected values already noted on the last-stage nodes are now used in conjunction with the related state-of-nature probabilities to calculate expected values for each of the second-last stage courses of action.

This procedure is followed stage-by-stage back to the first decision node in order to choose the initial course of action. The same procedure is used to determine the correct course of action at any intermediate stage, after the actual states of nature for the previously completed decision stages have become available.

Example 9.3 In Figure 9.11 a simple two-stage decision process is considered with two initial alternatives, a_{11} and a_{12}:

a_{11}: install water pipeline with 'small' capacity;
a_{12}: install 'large' capacity pipeline.

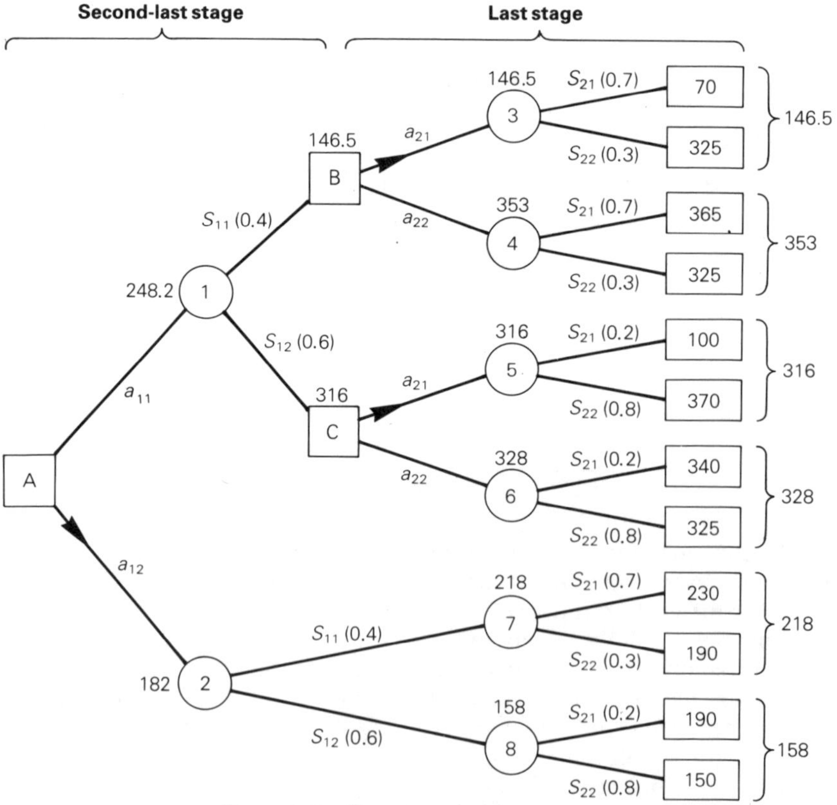

Figure 9.11 Two-stage decision process.

In the first time period, the subsequent outcomes in either case are 'low' demand or 'high' demand, which may here be represented by states of nature S_{11} and S_{12}, respectively. Having observed the states of nature, it is possible in the second time period to follow up action a_{11} with a second-stage decision, namely to do nothing (a_{21}) or to construct a second parallel pipeline to increase the capacity (a_{22}). This decision is again followed by two possible states of nature S_{21} and S_{22}, i.e. low or high demand.

Although the second stage of the process involving a_{21} could well be followed by subsequent stages, we consider in this simple example only a two-stage process. Note that there are no second-stage actions to be considered if the first-stage decision is to install a large-capacity pipeline (a_{12}).

The costs shown at the tips of the branches in Figure 9.11 take account of lost opportunity, as for example when the decision to install a large pipeline is followed by low usage. In the first stage of the process the probabilities of the states of nature are 0.4 and 0.6 for low demand and high demand, respectively. However, in the second stage, the probabilities are 0.2 and 0.8 for low demand and high demand, in the case where high demand has previously been experienced in the first stage. In the case of low demand in the first stage, the second-stage probabilities are 0.7 for low demand and 0.3 for high demand. The calculations are carried out in terms of costs, so that the negative signs can be dropped.

In order to analyse the problem, the expected cost is first calculated at node 3, as 0.7 (70) + 0.3 (325) = 146.5. Expected costs at nodes 3 to 8 are shown. The expected value at node 2 (which is also the expected cost of action a_{12}) is the weighted sum of the values at nodes 7 and 8, i.e. 0.4 (218) + 0.6 (158) = 182. At node B the expected cost is the lower of the values at nodes 3 and 4, i.e. 146.5. Likewise at C the expected cost is equal to that for the preferred action a_{21}, i.e. 316 at node 5. The preferred actions are indicated in Figure 9.11 by arrows emanating from the decision nodes. Finally, the expected cost at node 1 is obtained from those at B and C as 0.4 (146.5) + 0.6 (316) = 248.2. This is the expected value of action a_{11}, which is larger than for a_{12}.

The analysis in Figure 9.11 shows that, with the data given, action a_{12} is the better at Stage 1.

When the decision process has been analysed by the last-step-first procedure outlined above, the appropriate course of action is determined for each stage of the process, irrespective of the previous decisions and previous outcomes. Consider for example that action a_{11} is chosen in Figure 9.11, despite the outcome of the analysis. Although it will not be known in advance whether node B or node C will result, the appropriate course of action can be read directly from the decision tree for each possibility. These are indicated in the tree by the arrows leaving the (square) decision nodes.

Multi-stage decision processes can arise for which the decision tree becomes a very complex network. In cases where the multi-stage process has certainty of outcome, various network optimization procedures such as linear programming and dynamic programming can be used to obtain a solution. These techniques are dealt with in Chapters 7 and 8, although not specifically in relation to decision analysis.

9.5 Bayes' theorem

When additional information is obtained by experimentation or analysis, the question arises as to how it can best be incorporated into the decision process. We are concerned here with decisions with risk of outcome in which new information allows improved estimates to be made of the probabilities p_{ij}.

Improved estimates of the probabilities are obtained by an application of Bayes' theorem, which is concerned with the conditional probabilities of two related events. We consider two events, say X and E, which are interrelated so that the occurrence of E affects the probability of occurrence of X. The conditional probability of occurrence of X, given that E has occurred, is expressed as $P(X|E)$. Bayes' theorem states that this conditional probability may be expressed as follows:

$$P(X|E) = \frac{P(E|X) \, .P(X)}{P(E)} \qquad (9.4)$$

The validity of the theorem may be demonstrated with a Venn diagram (Figure (9.12)), in which areas x and e represent events X and E and the common area z represents the joint occurrence of X and E. The universal event is u, so that we have

$$\begin{aligned} P(X) &= x/u \\ P(E) &= e/u \\ P(X|E) &= z/e \\ P(E|X) &= z/x \end{aligned}$$

and we can therefore write

$$\frac{P(E|X) \, .P(X)}{P(E)} = \frac{z}{x} \cdot \frac{x}{u} \cdot \frac{u}{e} = \frac{z}{e}$$

showing that the expression reduces to z/e and hence $P(X|E)$.

To apply the theorem to the present decision situation we consider an exhaustive and mutually exclusive set of events $X_1, \ldots, X_j, \ldots, X_n$ which might be the possible outcomes of an action a, or alternatively the states of nature in a decision situation. The initial estimates of the probabilities are $P(X_1), \ldots, P(X_j), \ldots, P(X_n)$. We now imagine an experiment which leads to an observed event E. The Venn diagram in Figure 9.13 represents the situation.

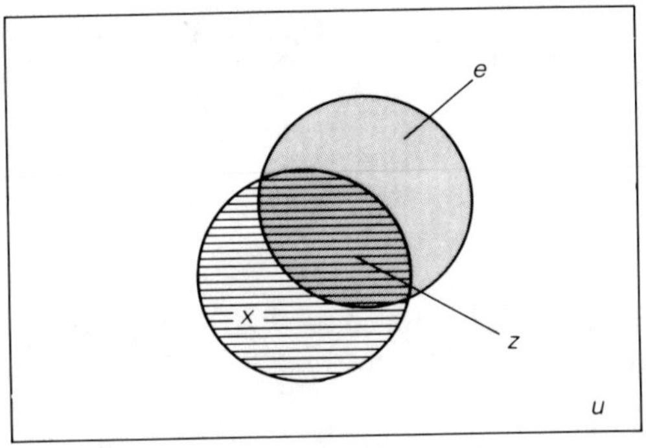

Figure 9.12 Venn diagram for events X and E.

Given that event E occurs, the improved estimate of the probability of occurrence of X_j, expressed as $P(X_j|E)$, is given by Bayes' theorem as

$$P(X_j|E) = \frac{P(E|X_j)\ P(X_j)}{P(E)} \tag{9.5}$$

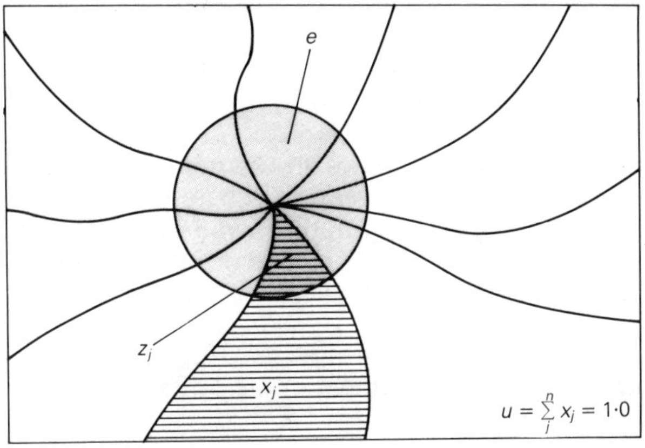

Figure 9.13 Venn diagram for events X_j and E.

In this instance, since the X_j are mutually exclusive and exhaustive, we have

$$P(E) = \sum_{j=1}^{n} [P(E|X_j)\ P(X_j)] \tag{9.6}$$

and the expression for $P(X_j|E)$ becomes

$$P(X_j|E) = \frac{P(E|X_j)\ P(X_j)}{\sum_{j=1}^{n} [P(E|X_j)\ P(X_j)]} \tag{9.7}$$

The validity of these equations can also be demonstrated by a Venn diagram as in Figure 9.13. For example, the right-hand side of Equation (9.6) becomes

$$\Sigma P(E|X_j)\ P(X_j) = \sum_j \frac{z_j}{x_j} \cdot \frac{x_j}{u} = \frac{1}{u}\sum_j z_j$$

which is $P(E)$. With this substituted for $P(E)$, and referring to Figure 9.13, the right-hand side of Equation (9.5) becomes

$$\frac{P(E|X_j)\ P(X_j)}{P(E)} = \frac{z_j}{x_j} \cdot \frac{x_j}{u} \cdot \frac{u}{\Sigma z_j}$$

$$= \frac{z_j}{\sum_j z_j}$$

which is $P(X_j|E)$.

Physically, we interpret $P(X_j)$ as the 'prior' probability of event X_j, i.e. the estimated probability of X_j before the experiment is carried out, while $P(E|X_j)$ is the estimated probability that the experimental event E would be obtained if X_j were the actual outcome. The improved, 'posterior' probability of X_j, given that E occurs, is $P(X_j|E)$.

Example 9.4 To illustrate the use of the theorem we consider a situation in which the construction of a prestressed concrete girder bridge has been unavoidably delayed, with the result that some rusting has occurred of the prestressing strand. The decision is whether or not to replace the strand because of a possible reduction in fatigue resistance due to local pitting in the strand surface.

The cost of replacement of the strand is \$20 000, whereas if fatigue failure were to occur in service, replacement of the bridge would cost \$750 000.

We first consider two courses of action:

a_1: replace the strand
a_2: do not replace the strand

The following outcomes are possible:

X_1: the strand is replaced and the bridge performs satisfactorily over its full design life
X_2: the strand is replaced but premature fatigue failure occurs
X_3: the strand is not replaced and premature fatigue failure does not occur
X_4: the strand is not replaced and premature fatigue failure does occur

With the estimated probabilities shown in Figure 9.14, strand replacement (a_1) is clearly the preferred action based on the criterion of expected monetary value.

However, a third option, a_3, is to postpone the decision until some fatigue tests can be carried out on samples of the strand. The cost of the tests is \$2000. To evaluate a_3 we consider two possible outcomes of the tests:

E_1: the tests indicate no significant reduction in fatigue life due to rusting
E_2: the tests indicate a significant reduction in fatigue life.

Considering first the situation if E_1 occurs, we make the following estimates in order to apply Bayes' theorem:

$$P(E_1|X_3) = 0.95$$

$$P(E_1|X_4) = 0.09$$

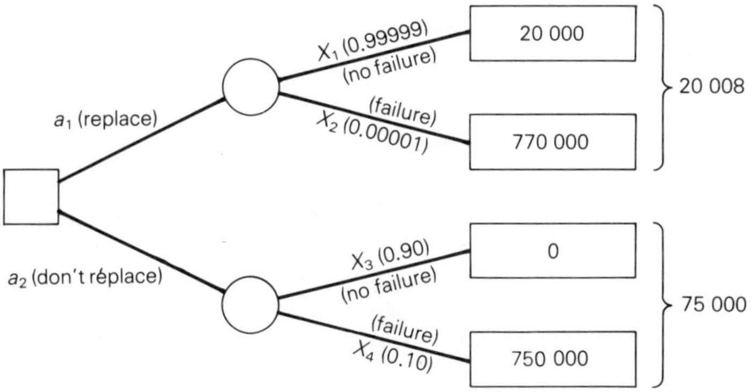

Figure 9.14 Initial decision analysis.

With the previously estimated values (Figure 9.14) $P(X_3) = 0.9$ and $P(X_4) = 0.1$, we obtain

$$P(E_1) = P(E_1|X_3).P(X_3)+P(E_1|X_4).P(X_4)$$

$$= 0.95 \times 0.9 + 0.09 \times 0.10 = 0.864$$

and hence

$$P(X_3|E_1) = \frac{0.95 \times 0.90}{0.864} = 0.99$$

$$P(X_4|E_1) = \frac{0.09 \times 0.10}{0.864} = 0.01$$

The results of the test have no implications regarding outcomes X_1 and X_2, and so the branch of the decision tree following action a_1 (replacement) is unchanged.

The portion of the decision tree following on from $a_3 - E_1$ is thus as shown in Figure 9.15. Action a_2 following on from E_1 has an expected cost of \$7500. With the cost of the tests included this becomes \$9500. It is clear that non-replacement is the preferred option.

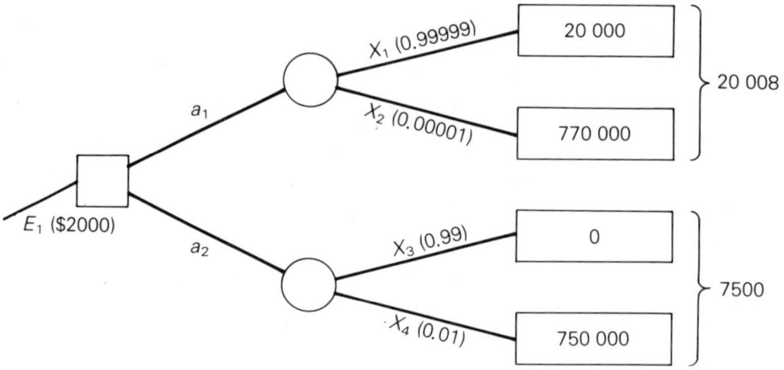

Figure 9.15 Decision branch following testing with outcome E_1.

On the other hand, if E_2 is the outcome of the tests, our estimates are

$$P(E_2|X_3) = 0.05$$

$$P(E_2|X_4) = 0.91$$

and with

$$P(E_2) = P(E_2|X_3).P(X_3)+P(E_2|X_4).P(X_4)$$

$$= 0.05 \times 0.90 + 0.91 \times 0.10$$

$$= 0.136$$

we obtain

$$P(X_3|E_2) = \frac{0.05 \times 0.90}{0.136} = 0.33$$

$$P(X_4|E_2) = \frac{0.91 \times 0.10}{0.136} = 0.67$$

The expected cost of non-replacement now becomes $0.67 \times (750\,000) =$

502 500 and clearly it is better to replace the strand, with expected cost \$20 008 (plus \$2000 for the cost of the tests), if the outcome is E_2.

In order finally to answer the question of whether it is worth while to choose action a_3 and test, in preference to a_1 and a_2, we need to look at the entire decision tree. In Figure 9.16 the original decision branches are shown, but for clarity the end branches and tips have been omitted and replaced by appropriate expected values.

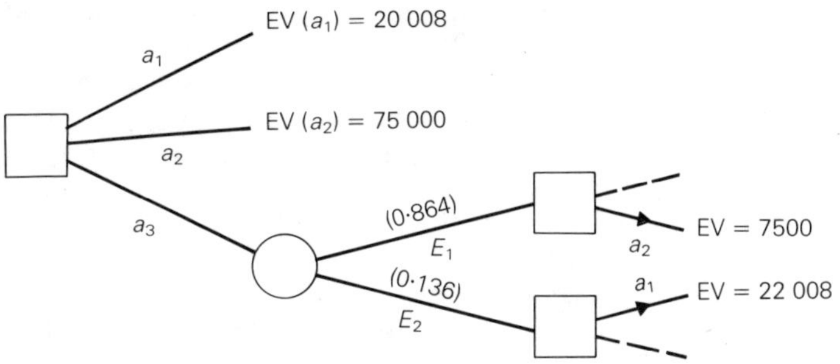

EV (a3) = 2000 + 0·864 (7500) + 0·136 (22 008) = 11 201
(this figure is a cost; the negative sign has been omitted throughout this example)

Figure 9.16 Evaluation of option a_3, testing.

The expected cost of a_3 is calculated as

$$\mathrm{EV}(a_3) = 0.864(7500) + 0.136(20\,008) + 2000$$

$$= 11\,201$$

and this is the best course of action.

9.6 Decision-making with competition

In some circumstances, the decision-maker has to take account of other persons who may try to influence the outcome of a decision, usually unfavourably. Business decisions frequently fall into this category; engineering decisions also, although less frequently, as for example when a construction company is deciding how to tender for an engineering project, knowing that other companies are also likely to tender.

Game theory is a relatively new branch of mathematics which deals with competitive decision-making. In principle, the concepts apply to many competitive situations such as occur in games of strategy, business, economic behaviour, and even tactical warfare. In fact, the real-world situations are usually too complex to allow direct application of the models which have been developed. Nevertheless, a good insight into decision-making in the face of competition can be obtained from a study of the elements of game theory. In this section, we take a brief look at some of the simplest aspects of game theory.

It is useful to distinguish between **strategic** and **non-strategic** games. Non-strategic games do not depend on any skills and do not require intelligent decisions to be made by the players. The outcomes of a non-strategic game depend purely on chance, and can be analysed with the help of probability and statistics. Examples of non-strategic games include board games such as snakes and ladders, and betting on the outcome of a tossed coin or a die.

A strategic game involves two or more players who act rationally to promote their own interest. Although the outcome of the game is influenced by the actions of the players, chance may also play a part in determining the outcome.

The simple one-stage decision with risk of outcome, which has been discussed in Section 9.2 in terms of courses of action and states of nature, could be regarded as a primitive two-person game played between the decision-maker and Nature. A 'play' of the game consists of the decision-maker choosing a 'move' i.e. one course of action from a number of possible actions. Nature responds by choosing a 'move', i.e. one state of nature from the range of possible states. The two moves together determine the outcome of the game. In this example, Nature tends to act in a disinterested manner, rather than malevolently.

In the normal strategic game the objective of each player is to win as much as possible, or at least to lose as little as possible. The outcome of a game requires that one or more players win, and the rest lose. A zero-sum game is one in which the sum of the winnings is equal to the sum of the losses.

TWO-PERSON ZERO-SUM GAME

A two-person zero-sum game involves two players A and B. Player A has a number of possible moves, which are denoted as $a_1, \ldots, a_i, \ldots, a_m$. Player B likewise has a number of possible moves, $b_1, \ldots, b_j, \ldots, b_n$. A play occurs when the players each choose a move, not knowing at the time the move chosen by the opponent. To each possible pair of moves (a_i, b_j) there is an outcome O_{ij} which by convention involves a payoff u_{ij} from B to A. A negative payoff means that B in fact wins from A. The essentials of the game are summarized in the payoff matrix, as in Table 9.5 which applies to a specific example.

The problem in Table 9.5 is to determine the best strategies for A and B. We consider first the way A might reason. Initially, an attempt might be made to obtain the maximum possible payoff, which in Table 9.5 is

$$\max_{ij}(u_{ij}) = u_{32} = +13$$

and would require move a_3. However, recognizing that B is intelligent and unlikely to choose move b_2, A can reasonably conclude that it is not feasible to achieve the maximum possible payoff.

Table 9.5 Two-person zero-sum game. Payoffs shown are from B to A.

(a) Payoff matrix

	b_1	b_2	b_3
a_1	−1	+4	−3
a_2	+1	+2	+2
a_3	−2	+13	+1

(b) Maximin strategy for A

Payoff matrix		B b_1	b_2	b_3	$\min_j(u_{ij})$	$\max_i[\min_j(u_{ij})]$
	a_1	−1	+4	−3	−3	
A	a_2	**+1**	+2	+2	+1	+1 ($i = 2$; $j = 1$)
	a_3	−2	+13	+1	−2	

(c) Minimax strategy for B

		B b_1	b_2	b_3	
	a_1	−1	+4	−3	
A	a_2	**+1**	+2	+2	
	a_3	−2	+13	+1	
	$\max_i(u_{ij})$	+1	+13	+2	
	$\min_j[\max_i(u_{ij})]$	+1			($j = 1$; $i = 2$)

A reasonable alternative for A is to consider the minimum payoff from each possible move, and choose the move which maximizes the minimum payoff. The minimum payoffs, $\min_j(u_{ij})$, are listed in the column in Table 9.5(b) next to the payoff matrix. The result of this maximin strategy is

$$\max_i \, [\min_j(u_{ij})] = +\,1 \; (i{=}2;\, j{=}1)$$

which requires move a_2.

From B's point of view, the initial attempt might be to minimize the payoff,

$$\min_{ij} (u_{ij}) = -3$$

which would require move b_3. Recognizing that A is intelligent and unlikely to choose a_1, B looks for an alternative criterion, and as with A, has to be content with less than the best outcome. B is left with the strategy of considering the maximum payoff from each move, as in Table 9.5(c), and minimizing the maximum payoff. This gives

$$\min_j[\max_i(u_{ij})] = +1 \ (j=1; i=2)$$

and requires move b_1.

In this particular game the point (i=2, j=1) represents the most satisfactory result for both players. This point is said to be a **saddlepoint**, for which

$$\min_j[\max_i(u_{ij})] = \max_i[\min_j(u_{ij})]$$

and $u_{21} = +1$ is the value of the game. Even if the game is played repeatedly, there is no advantage for A or B in playing other than the moves already identified. A sudden change in strategy is likely to lead to a worse result for both players.

It is clear that this particular game is unfair to B, because A can always win. The decision criteria are:

For A: maximize the minimum payoff
For B: minimize the maximum payoff

The strategies which follow from the criteria in each case involve only one move, and are shown as follows:

For B: [1, 0, 0]

For A: $\begin{bmatrix} 0 \\ 1 \\ 0 \end{bmatrix}$

Not all games have saddlepoints. In the simple example in Table 9.6, we have

$$\max_i[\min_j(u_{ij})] = +2 \ (i=1, j=1)$$

$$\min_j[\max_i(u_{ij})] = +3 \ (j=1, i=2)$$

Table 9.6 Game without a saddlepoint.

	B: b_1	B: b_2	$\min_j(u_{ij})$	$\max_i[\min_j(u_{ij})]$
a_1	**+2**	+5	+2	+2
a_2	**+3**	−5	−5	
$\max_i(u_{ij})$	+3	+5		
$\min_j[\max_i(u_{ij})]$	+3			

Strategy for B $[q, (1-q)]$ $\quad 0 \leqslant q \leqslant 1.0$

Strategy for A $\begin{bmatrix} p \\ (1-p) \end{bmatrix}$

$0 \leqslant p \leqslant 1.0$

If the game is to be played once only, it is not clear which would be the best move for each player. To maximize the minimum payoff, A must play a_1; to minimize the maximum payoff B must play b_1. However, in the expectation that B will play b_1, A might be tempted to play a_2, thereby increasing the payoff to +3 but running the risk of the payoff falling to −5. Likewise, in attempting to double-guess A's strategy B might be tempted to play b_2, to achieve a payoff of −5, but with the risk of the payoff increasing to +5.

If the game is to be repeated a large number of times, A can attempt a mixed strategy by unexpectedly playing a_2 now and then in an attempt to improve the long-term payoff. B can respond by playing b_2 occasionally. A mixed strategy can be found for each player which involves choosing each play a definite proportion of the time, but in a random and, from the viewpoint of the opponent, unpredictable manner. The strategy for A is represented by the probabilities p and $(1-p)$, where p is the probability or relative frequency of choosing move a_1, and $(1-p)$ is that for move a_2. For B, the strategy is represented by q and $(1-q)$.

We have:
For B: $[q,(1-q)]$

For A: $\begin{bmatrix} p \\ (1-p) \end{bmatrix}$

Values of p and q can be determined on the basis of maximizing the minimum expected payoff for A and, for B, minimizing the maximum expected payoff. For a general 2×2 game (Table 9.7) the relative frequency of payoff u_{11} is $pq\,u_{11}$. The total expected value of the game depends on p and q and is written as

$$EV(p,q) = pq\, u_{11} + p(1-q)u_{12} + (1-p)q\, u_{21} + (1-p)(1-q)u_{22} \qquad (9.8)$$

Table 9.7 General 2×2 game.

			B: b_1	B: b_2
			q	$(1-q)$
A	a_1	p	u_{11}	u_{12}
	a_2	$(1-p)$	u_{21}	u_{22}

$$EV = pq\, u_{11} + p(1-q)\, u_{12} + (1-p)q\, u_{21} + (1-p)\,(1-q)\, u_{22}$$

For any chosen value of p, EV is a linear function of q. However, A can choose a specific value p^* such that EV is independent of B's strategy, i.e. EV appears as a horizontal line when plotted against q. To find p^* we set the differential of EV, with respect to q, equal to zero:

$$\frac{d\,(EV)}{dq} = p^*\,[u_{11} - u_{12} - u_{21} + u_{22}] + u_{21} - u_{22} = 0$$

and hence

$$p^* = \frac{u_{22} - u_{21}}{u_{11} - u_{12} - u_{21} + u_{22}} \qquad (9.9)$$

If A uses p^*, the EV is fixed at

$$EV(p^*,q) = \frac{u_{11}\,u_{22} - u_{12}\,u_{21}}{u_{11} - u_{12} - u_{21} + u_{22}} \qquad (9.10)$$

On the other hand, B can also choose a strategy q^* such that the EV is independent of A's strategy. Setting the differential of EV, with respect to q, to zero, we obtain

$$q^* = \frac{u_{22} - u_{12}}{u_{11} - u_{12} - u_{21} + u_{22}} \qquad (9.11)$$

for which the EV is fixed at

$$EV(p,q^*) = \frac{u_{11}\,u_{22} - u_{12}\,u_{21}}{u_{11} - u_{12} - u_{21} + u_{22}}$$

This expression for EV is the same as that in Equation (9.10), and it can be shown that p^* and q^* represent optimal strategies.

For the numerical case shown in Table 9.6, the values of p^* and q^* are:

$$p^* = \frac{-5 - 3}{2 - 5 - 3 - 5} = \frac{8}{11}$$

$$q^* = \frac{-5 - 5}{2 - 5 - 3 - 5} = \frac{10}{11}$$

and the expected value of the optimal game is

$$\mathrm{EV}^* = \frac{2 \times (-5) - 5 \times 3}{2 - 5 - 3 - 5} = \frac{25}{11}$$

More generally, algebraic analysis can be carried out for $m \times n$ games. It is found that such analysis has very close parallels with linear programming and indeed that a two-person zero-sum game can be analysed as an LP problem (Taha 1982).

Although a detailed study of game theory is beyond the scope of the present text, one general point is of interest. It will be clear that the plays in any mixed strategy have to be conducted randomly. If B finds any pattern in the sequence in which A chooses a_1 and a_2, advantage can be taken of this additional information to improve the result from B's point of view. It is for this reason that an opponent in a competitive situation, whether a strategic game, business, or even warfare, should never be allowed to know of your next move, even if your overall strategy can be predicted. On the other hand, it is of great advantage to determine your opponent's overall strategy and, if possible, next moves. Spies are invaluable in giving inside information on moves and strategies, not only in wartime but also in industrial and commercial contexts.

OTHER FORMS OF GAMES

We have considered only a simple game which is highly idealized when compared with real engineering decision situations. Other, more complex games have been studied theoretically but these still do not provide methods of analysis which are of direct practical use. Nevertheless, they provide an insight on real situations.

A simple extension of the zero-sum game to include three players leads to a great increase in complexity. With three players A, B, and C, the strategy of forming a coalition with one player against the third player becomes an important option. For example, A and B may join forces, temporarily, to win jointly from C. In such a circumstance, a strategy for C is to try to break the coalition and then to form a coalition with either A or B. This is possible because the best strategy for A is not necessarily to remain in the current coalition with B. For example, if the coalition has been successful and C has incurred great losses, there may well be greater winnings to be obtained from B, by taking C as an ally.

It is also possible to contemplate games which are not zero-sum, i.e. games for which the total losses do not have to be equal to the winnings. Such games are difficult to study mathematically and are not well understood. Optimal strategies can only be obtained in complex non-zero-sum games by trial and error, or by simulation and search.

9.7 Contingency planning

It will be appreciated from previous discussions that whenever decision-making is undertaken with less than certainty of outcome, it is possible that the best course of action may lead to unwanted and indeed unfortunate outcomes. This is an inherent aspect of the uncertain, real world.

In planning for an engineering project, and in engineering decision-making generally, it is therefore very important to study carefully all the possible outcomes, even the unlikely ones, and where possible and appropriate to make provision for their occurrence.

For example, in the design of a commercial building or a bridge it is impossible to guarantee absolute safety against the largest conceivable earthquake which might occur. Even design against a large, but unlikely, earthquake can prove to be exorbitant in cost. On the other hand, various detailing techniques can be employed which, with little additional expense, greatly improve the ductility and robustness of the structure. While not assuring that the structure remains undamaged, good ductility and robustness help prevent catastrophic collapse and thus decrease the danger of loss of life in a catastrophe. Other detailing techniques can be used which would greatly reduce the spread of fire in a building, while not completely eliminating the possibility of an outbreak.

As part of the planning and design for any major engineering project, contingency plans have to be developed for coping with natural disasters and non-functioning or failure of various parts of the system. Contingency planning together with follow-up development work clearly adds to the overall cost of the project. How far such work should be carried out is yet another difficult engineering decision.

References

Asplund, S. O. 1968. The risk of failure, *Structural Engineer* **36,** No. 8, 268–70.

deNeufville, R. & J. H. Stafford 1971. *Systems analysis for engineers and managers*. New York: McGraw-Hill.

McDonald, J. D. 1950. *Strategy in poker, business and war*. New York: Norton Paperbacks.

McKinsey, J. C. C. 1952. *Introduction to the theory of games*. New York: McGraw-Hill.

Meek, R. L. 1971. *Figuring out Society*. London: Penguin.

Pratt, J. W., H. Raiffa & R. Schlaiffer 1965. *Introduction to statistical decision theory*. New York: McGraw-Hill.

Raiffa, H. 1970. *Decision analysis, introductory lectures on choices under uncertainty*. Reading: Addison-Wesley.
Taha, H. A. 1982. *Operations Research*, 3rd edn. New York: Macmillan.
von Neumann, J. & O. Morgenstern 1947. *Theory of games and economic behavior*. Princeton: Princeton University Press.
Williams, J. D. 1966. *The compleat strategist*. New York: McGraw-Hill.

Problems

9.1 In the decision tree shown in Figure 9.17 the symbols have the following meanings:

a_1: you choose to pay $\$X$ to take part in a 'lottery' with two possible outcomes, O_1 and O_2
a_2: you choose not to take part in the lottery
O_1: you win $\$W$, the probability of this event being p^*
O_2: you win nothing

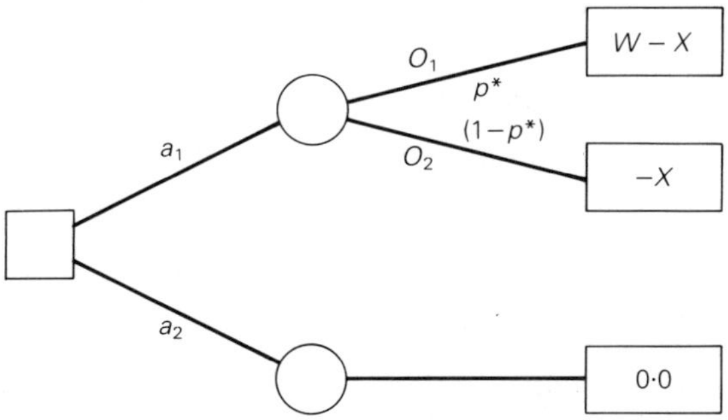

Figure 9.17 Decision tree for Problem 9.1.

(a) Choose a sequence of values of X, and for W = \$10 determine the limiting values of p^* at which you personally decide to take action a_1. Tabulate your results to give values of p^* and X/W versus X.
(b) Repeat the exercise, this time for W = \$1000.
(c) Plot p^* versus X/W for (a) and (b) and hence compare any conditions of risk aversion and risk acceptance for the two cases.
(d) Draw up personal utility functions for the two cases of gains and losses in the order of \$10 and \$1000, respectively.

9.2 A government department has designed a large reinforced concrete multi-span bridge and has called tenders for its construction.

You must decide on one of the three following courses of action, on behalf of your employer:

(a) do not tender;
(b) prepare and submit a tender, based on the reinforced concrete design;

(c) prepare a new design using a cheaper form of prestressed concrete construction and then tender for the job on the basis of this new design.

You estimate that, if you get the contract for the reinforced concrete design, there is a 70 per cent chance of making a $60 000 profit, a 15 per cent chance of breaking even, and a 15 per cent chance of making a loss of $300 000. The cost of preparing the tender is $20 000, and this cost has been taken into account in the above profit-and-loss and break-even estimates. The chances of winning the contract for the reinforced concrete design are good: about 70 per cent.

Although the chances of winning the contract on the basis of the prestressed concrete design are much lower, about 40 per cent, you stand to make a much larger profit. You estimate that, if you get the contract, you would have a 70 per cent chance of making a profit of $1 500 000. There is a 30 per cent chance of losing $300 000. The cost of preparing the tender in this case is $30 000, but there is the added cost of $200 000 for the preparation of the new design. Both of these costs have been included in the profit-and-loss estimates given.

Make any calculations you consider appropriate, state any assumptions you decide to make, and hence indicate your decision.

What is the expected value of your decision?

9.3 A reinforced concrete grain silo complex is showing signs of structural defects which are probably due to design errors. Following a structural assessment, three possible courses of action are being considered.

The first action consists of demolishing and rebuilding the silos at a cost of $1.95M. The second action requires strengthening of the structure at an estimated cost of $800 000. The third action is to do nothing, but to monitor the structure to detect any further deterioration. The cost of this action is negligible.

Calculations suggest that there is a real likelihood of collapse of the structure. There is a probability of collapse of 0.10 for the unstrengthened structure, but this reduces to 0.02 if strengthening is undertaken. The total losses from collapse would be in the order of $6M. If the structure is totally rebuilt the chance of failure is assumed to be zero.

If the structure is not strengthened there is a 30 per cent likelihood that further deterioration will lead to the need for repairs, which would cost $800 000, but which would be certain of success.

(a) Describe a set of circumstances in which it would be appropriate to use the expected monetary value criterion to determine the most appropriate action.
(b) Which is the preferred action according to the expected monetary value criterion?
(c) Describe a set of circumstances for which it would not be appropriate to use this criterion. How could the decision problem be analysed in this case?

9.4 The director of a firm of consulting engineers must decide whether or not to take out professional indemnity insurance for the coming year. The premium is \$10 000.

She estimates that there is a probability of 0.02 that the firm will be sued in the coming year for professional liability due to unsatisfactory performance of any building designed by the firm. The total damage involved will be either \$100 000 or \$1 000 000 with equal probability. Professional indemnity insurance will cover all such costs.

If no insurance is held and the firm is sued, the director must choose between an out-of-court settlement costing 50% of the damage or contesting the case in court. In the latter situation, she estimates there is a 60% chance of winning the case if the damage is \$100 000 and an 80% chance if the damage is \$1 000 000. If she contests the case and loses she must pay the total damages plus legal fees estimated to be 20% of the damage. If she contests the case and wins she must pay only the legal fees.

(a) Construct a decision tree for the problem showing all possible actions, states and outcomes.
(b) The director has a utility function given by:

$$u(X) = \begin{cases} 1 + (0.5 \times 10^{-6})X & \text{for } -600\,000 \leqslant X \leqslant 0 \\ 1.4 + (1.167 \times 10^{-6})X & \text{for } -1\,200\,000 \leqslant X \leqslant -600\,000 \end{cases}$$

where X is the monetary outcome in dollars

Identify whether professional indemnity insurance should be purchased taking into account the director's utility function.

9.5 A construction company must decide on the best plan to follow for the installation of a pipeline. Plan A is a minimum-time approach which will require 3 months for the installation at a cost of \$200 000. Plan B is to follow normal construction procedures; the time required for installation is 4 months and the cost is \$160 000. Plan C involves a slower approach to installation; the time required is 5 months and the cost is \$150 000.

The installation must be undertaken during winter and additional costs will be incurred if bad weather occurs. Essentially, only three weather conditions need to be distinguished: a dry winter; an average winter; and a wet winter. Additional construction costs for each plan and set of weather conditions are given in Table 9.8.

Assuming that the probability of occurrence of each weather type cannot be estimated from existing data, which plan should be chosen under each of the following criteria:

(a) pessimistic
(b) optimistic
(c) minimum regret
(d) Laplace

What do you think is the best construction plan and why?

Table 9.8 Additional construction costs.

	Construction plan		
	A	B	C
Dry winter	0	0	0
Average winter	$15 000	$60 000	$90 000
Wet winter	$20 000	$90 000	$120 000

Index